Praise for
Climate, Psychology, and Change

"The climate crisis is the greatest physical crisis we've ever faced as a species, but maybe the greatest metaphysical crisis too: as our assumptions about the present and future, about safety and risk, about individuality and solidarity are upended, it is bound to be both psychologically hard and psychologically rich—and as these intriguing essays make clear, some of the finest minds in the world are thinking through the problems and arriving at powerful answers."

　　—BILL MCKIBBEN, author, environmentalist, educator, activist, and
　　　founder of Third Act

"Steffi Bednarek has curated a collection of texts that asks extremely important questions about what psychotherapy might look like beyond its mainstream individualistic, anthropocentric, and Western/colonial frameworks. Without providing definitive answers, this book invites the readers to consider how the focus on individual mental health and wellbeing is preventing us from recognizing how our current inner cognitive, affective, and relational infrastructures are tied to the collapsing infrastructures that surround us. *Climate, Psychology, and Change* is an invaluable companion to those interested in who we can be once we process the difficult lessons of modernity dying within and around us."

　　—VANESSA MACHADO DE OLIVEIRA, author of *Hospicing Modernity*

"*Climate, Psychology, and Change* is an outstanding update of the ecopsychology movement. A broad range of perspectives are brought together in this collection of marvelous essays showing that the human soul cannot be healthy on a sick planet. The healing of the *anima mundi* and the healing of the human soul depend on each other. It is more than a book; it is a treasure of radical ideas and profound insights. It is a book of wisdom! Steffi Bednarek has woven a garland of great thoughts which will inspire the reader to look at the world and see it as an interconnected whole."

　　—SATISH KUMAR, founder of Schumacher College and editor emeritus of
　　　Resurgence & Ecologist

"Climate change is a systemic problem with geophysical, technological, economic, political, and ethical dimensions, among others. The in-depth exploration of our climate crisis from a mental health perspective offered in this book will be an important contribution to an urgently needed dialogue."

—FRITJOF CAPRA, author of *The Web of Life* and coauthor of *The Systems View of Life*

"*Climate, Psychology, and Change* is an exquisitely crafted deep dive into the profound uncertainties we must face in these times of climate breakdown. The editor's intention to 'bring regenerative disturbance to the commons of our profession' is well-fulfilled. This is a prophetic book, necessarily disturbing, articulating many necessary paradoxes. It is a catalytic gift to the psyche professions and beyond."

—JUDITH ANDERSON, Jungian analytical psychotherapist, chair of board of trustees for the Climate Psychology Alliance

"Nothing is what it seems, the symptom is not the problem, psychology is not just psychology. There is a necessary blurring that brings transcontextual combining into every moment of life. The response is not a strategy, but rather a shifted ecology of perception. As this era unravels the stitchery of so many destructive illusions woven tightly into the last several centuries, new questions are surfacing. The familiar is a trap of traps wrapped and soaked in separations that perpetuate the existing habits over and over again. There is so much possibility just waiting—but it looks nothing like it used to. This beautiful book leaves nothing behind."

—NORA BATESON, filmmaker, author, and founder of Warm Data

"This powerful collection explores how psychotherapy can help us face the unfolding reality of climate change, focusing on a wide range of crucial questions including how we can build communal containers to help us hold our grief, rage, and fear, and how we can learn to stay with the unknown, reacting to the world as it emerges rather than pushing for premature solutions. An extraordinary and hugely timely book."

—REBECCA HENDERSON, John and Natty McArthur University professor at Harvard University

"As we ponder the climate crisis—how we could have let ourselves get to this point and what we can do about it—the voices of mental health professionals have often gone missing. Until now. For those seeking new ways to understand and take action, there are answers. In *Climate, Psychology, and Change*, renowned authors share their knowledge, passion, and experience, bringing the psychological components and complexities of our climate predicament out of the shadows. And as important as numbers are, even more telling is their recounting of the root of the climate crisis—the crisis of the human spirit. This book tells that story, and more."

—LISE VAN SUSTEREN, MD, psychiatrist, specialist in climate and mental health, and founder of the Climate Psychiatry Alliance

"*Climate, Psychology, and Change* is an urgent and necessary response to the most critical emergent questions of our time. Steffi Bednarek and the contributors offer a bold vision that reimagines the role of psychotherapy and widens the ways we think about what it means to be human. Not only does the book equip us with the skills to help clients navigate climate anxiety, eco-distress, and disruption, but it asks us to stretch our imagination beyond the assumptions of an outdated worldview. The authors move our focus from the care for the individual to practices that are grounded in collective action. This beautiful book illustrates how to meet this moment with care and grace, even as we look toward uncertain futures."

—BRITT WRAY, PhD, director of CIRCLE at Stanford Psychiatry and author of *Generation Dread*

"For well over a century, psychology has tried to help us better adjust to what is. In this book, a refreshing diversity of psychologies asks us to realize that being well adjusted to 'what is' is in fact deadly. If before we were challenged to grow up, this book offers the container that will help us break down. It shows us that healing our personal traumas was never a big enough lens for the incipient pain inside and around us. It reads to me as if—in these pages—psychology itself comes into its own maturity; as if it could be the true elder, the trustworthy guide, that so much of humanity needs to make it through this dangerous, painful passage toward a saner, wiser world."

—SUSANNE MOSER, PhD, founder of The Adaptive Mind Project and director of Susanne Moser Research & Consulting

"The climate crisis is not just 'out there'; it is also a crisis of the disembedded and encapsulated modern self. Deploying contributors from across the world, Steffi Bednarek has assembled not only a multi-perspectival critique of this self but also a ferment of new possibilities for the practice of being therapist, citizen, and human."

—PAUL HOGGETT, cofounder of the Climate Psychology Alliance

"What an exciting book! Its very forms speak to what is different about it: for only genuine dialogue, within a broader container of the more-than-human, can indicate a path forward from the wrecked civilization whose psyches are now seeking to change.

This book is not a tedious chorus of agreement, but an emergent dialogue of deep insights and ideas that criss-cross one another. So refreshing!

What this book evinces is both a subtlety and a seriousness about helping climate distress to enter and alter psychotherapy. The difficult eco-emotions are an invitation to us all: to refind our collective voice, to become who we are, and to manifest as deep determination the pain we feel for the Earth and for each other. This book issues a wonderfully wild invitation."

—RUPERT READ, professor at the University of East Anglia and codirector of the Climate Majority Project

Climate,
Psychology,
and Change

Climate, Psychology, and Change

Reimagining Psychotherapy in an Era of Global Disruption and Climate Anxiety

EDITED BY STEFFI BEDNAREK

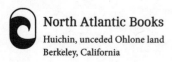

North Atlantic Books
Huichin, unceded Ohlone land
Berkeley, California

Published by
North Atlantic Books
Huichin, unceded Ohlone land
Berkeley, California

Cover photo © Markus Spiske via Unsplash
Cover design by Aashim Raj
Book design by Happenstance Type-O-Rama

Printed in the United States of America

Climate, Psychology, and Change: Reimagining Psychotherapy in an Era of Global Disruption and Climate Anxiety is sponsored and published by North Atlantic Books, an educational nonprofit based in the unceded Ohlone land Huichin (Berkeley, CA) that collaborates with partners to develop cross-cultural perspectives; nurture holistic views of art, science, the humanities, and healing; and seed personal and global transformation by publishing work on the relationship of body, spirit, and nature.

MEDICAL DISCLAIMER: The following information is intended for general information purposes only. Individuals should always see their health care provider before administering any suggestions made in this book. Any application of the material set forth in the following pages is at the reader's discretion and is their sole responsibility.

North Atlantic Books's publications are distributed to the US trade and internationally by Penguin Random House Publisher Services. For further information, visit our website at www.northatlanticbooks.com.

Library of Congress Cataloging-in-Publication Data

Names: Bednarek, Steffi, 1970– editor.
Title: Climate, psychology, and change : New perspectives on psychotherapy in an Era of global disruption and climate anxiety/ edited by Steffi Bednarek.
Description: Berkeley, California : North Atlantic Books, [2024] | Includes bibliographical references and index. | Summary: "28 leading psychologists, therapists, and mental-health healers reflect on the potential-and necessity-of adapting clinical care in response to the climate crisis"— Provided by publisher.
Identifiers: LCCN 2023042499 (print) | LCCN 2023042500 (ebook) | ISBN 9798889840817 (paperback) | ISBN 9798889840824 (epub)
Subjects: LCSH: Climatic changes—Psychological aspects. | Mental health. | Psychotherapy.
Classification: LCC BF353.5.C55 C554 2024 (print) | LCC BF353.5.C55 (ebook) | DDC 155.9/15—dc23/eng/20240126
LC record available at https://lccn.loc.gov/2023042499
LC ebook record available at https://lccn.loc.gov/2023042500

1 2 3 4 5 6 7 8 9 KPC 29 28 27 26 25 24

North Atlantic Books is committed to the protection of our environment. We print on recycled paper whenever possible and partner with printers who strive to use environmentally responsible practices.

For all of us

May we weave back together
what has been torn apart

and create the conditions that
allow life to thrive

Everywhere

Contents

Foreword

Thomas Hübl, PhD

"The familiar is dying."

The theme that weaves itself through this illuminating book's opening dialogue is a stepping stone into a threshold. As the reader walks through, the dimensions of this threshold of change, dismemberment, fragmentation, and disintegration are revealed through a kaleidoscope of wisdom and professional expertise. We, as readers, are embarking on a journey that elucidates this precise cycle of death, when what falls away around us and within us are the necrotic tissues of the past—the "normal" that is actually a faint echo of a pulse, now barely perceptible to our ears.

As you read this book and enter into the communal conversation that permeates this prescient collection of essays thoughtfully edited by Steffi Bednarek, you will move through the perspectives of the individual, weave into the collective, and then return to your perspective as the individual reader. This way of reading introduces a new way of engaging in conversation around the climate—and the associated traumas—as we shape our reference points as both individuals and collectives. We move from the consulting room into the world, and we then bring the world into the consulting room. We understand the deleterious impact of colonialism on our lives as individuals, and on our planet. We navigate through our separate, enclosed spheres as individuals, then realize our interdependence with all creatures and the rest of the natural world. We come to understand the level of disembodiment that permeates our lives, then begin to learn to sense the connection to the ground below our feet. And we realize that we don't live *on the planet,* but *as the planet.*

In this way, *Climate, Psychology, and Change* becomes a gateway to illuminate the practice of healing as we perceive ourselves *through the lens of the planet itself.* If we could see through this larger lens, what would Earth tell us about her needs for healing? When we view Earth from outer space, our perspective widens as witnesses. From this vantage point, we can see Earth's beauty, but we also need to become receptive to the tremendous wounds and scars she holds. With our feet firmly grounded in her soil, we can see and experience the actual suffering.

We find the familiar on shaky ground, not only as our natural environment breaks down, fires and floods rage, and weather patterns disrupt and destroy, but also in the reverberating impacts on our global public health. Women, who make up the vast majority of people forcibly displaced by climate change, are exposed to greater risks for sexual violence. These seismic shifts jeopardize food security, water access, and our habitats. Oppression, racism, ableism, and other forms of polarization worsen in the face of the demise of many of the Earth's ecosystems.

Climate disruptions are landing on fractured landscapes that have been broken apart by unresolved collective traumas of the past. We have all been born into a world shaped by trauma. For many of us, this perception of a collectively traumatized world is accepted as "normal" and "the way the world is." I would say: *this is how life is when we are hurt.*

For some of us, the familiar dangles its empty promises of comfort, continuity, safety, and well-being. For others, the familiar is composed of oppression, inequality, polarization, and war. It's easy to be magnetized by the familiar, drawn into a false dichotomy that divides our psyches into "safety" or "insecurity," or we violently oppose it, protesting until we kill the familiar.

My work over the past twenty years has taught me that the source of any significant global crisis originates in the collectively traumatized space we currently inhabit as humanity. In this state, our modern societies have lost the ability to generate healthy feedback loops, which every living system needs to generate to maintain stability. In a living system, feedback loops facilitate a self-regulatory process, bringing the system into balance when equilibrium is threatened. This flow of information allows a system to adapt and change so it can move in relation to the larger system within which it is embedded. We see the effects of this loss of equilibrium throughout our climate crisis. This, I believe, is one way to understand the term "climate trauma."

Modernity's concept of the familiar is rooted in what is nonemergent and stagnant. When there is a state of trauma in the collective body, we cannot generate solutions from a place of emergence and creativity. When change is not possible, a crisis is set in motion.

Another way to understand climate trauma is to examine our mixed responses as humanity to the crisis. The disruption that has ensued manifests not only in the losses in biodiversity we observe, global warming, and the ongoing shifts in weather patterns, among other manifestations, but also as a derangement of Modernity to make impactful decisions for our planetary well-being. Those of us who are aware of the ensuing harm, including policy makers and governments, realize there is an actual urgency to form a concerted global response. However, the response of our industrialized, hegemonic leadership is fueled by hyperactivation and stress, which underlies the collective trauma, constituting the sand in the engine of our current immobilization. The first critical step is to slow down so that we may better formulate the appropriate, integrated response to this urgency. Only in slowing down—while consciously responding to the urgency—will we heal and integrate what underlies our current climate trauma. Also at play in our modern cultures are denial, numbness, and "absencing," as Otto Scharmer refers to it—relinquishing all responsibility for and ownership of the planet as our very own nature. On the other hand, there is despair and a prevalence of climate anxiety, which especially impacts young people.

It is in this complexity that we discover the pulse of the new, the awakening of all our senses and calls to respond, the *response-ability* that is the call to live, the call to thrive. The strong separation that we see in the world currently diminishes our capacity for global collaboration. Moving through this threshold of change, we are embarking upon co-creating a world based not on our collective wounds but on collective well-being that is rooted in our interdependence.

"The familiar is dying."

I believe these words represent the threshold into which we, as humanity, are walking. We might view this threshold as an opening to the Soul, a deepening of our lives, our paths, our journey, as that which is of the Soul, as psychotherapist Francis Weller beautifully elucidates in the opening conversation of this book, describing the trails that the Soul lays down in our individual, communal, cultural, and planetary lives. The familiarity of being an individual is dropping away. Only our conscious awareness can recognize the truth of this.

As we learn to see through the broken glass of trauma, we can engage in the global collaboration we need to solve this crisis.

The journey through this dying process, this initiation, is sparing no one. Through restoration, we don't return to the familiar, to a "normal," but to a future that is yet unknown. We begin to harvest new ways of being and new learning that we could never have imagined before. We experience the blessings of posttraumatic growth as we integrate these traumas. When we realize that the individual and the collective are interdependent, we access the healing power inherent in that flow of intelligence. Our creativity flourishes—including the resolve to commit to viable solutions—as we experience a collective liquefaction, an unfreezing of the old, a release of the familiar. To open this door, with a client, with a group, within community, is to fulfill our collective Soul's sacred relationship with the Earth, and with one another. As we turn the pages of *Climate, Psychology, and Change,* we cross into a frontier that is ripe with possibility, expansive in its vision, and rich in embodied wisdom.

About the Authors

Matthew Adams is a principal lecturer in psychology in the School of Humanities and Social Science, University of Brighton, UK. His most recent book is *Anthropocene Psychology: Being Human in a More-Than-Human World*.

Cash Ahenakew holds a Canada Research Chair in Indigenous Peoples' Well-Being. He is also an associate professor in the Department of Education at the University of British Columbia. Cash is Plains Cree and is a member of the Ahtahkakoop Cree Nation. His research is based in a commitment to the development of Indigenous theories, curriculum, pedagogies, and mixed methodologies. His work addresses the complexities at the interface between Indigenous and non-Indigenous knowledges, education, methodology, and ceremony.

Bayo Akomolafe is a philosopher, a psychotherapist, and founder of The Emergence Network. He has been visiting professor at Middlebury College and Sonoma State University in the United States, Simon Fraser University in Canada, and Schumacher College in the UK. Bayo was born in Nigeria and is a descendant of the Yoruba people. Today he lives in India with his wife and family. He is author of *These Wilds Beyond Our Fences: Letters to my Daughter on Humanity's Search for Home*.

Glenn A. Albrecht is an honorary associate in the School of Geo-Sciences, University of Sydney. He retired as a professor of sustainability at Murdoch University in 2014 but continues to work as an environmental philosopher. He is author of the book *Earth Emotions,* and he lives on Wonnarua land in New South Wales, Australia. He is best known for creating the concept of "solastalgia," or the lived experience of negative environmental change.

Vanessa Andreotti is dean of the Faculty of Education of the University of Victoria. She has been a professor in the Department of Education at the University of British Columbia and the interim director of the Peter Wall Institute for Advanced Studies. She is also a former Canada Research Chair in Race, Inequalities, and Global Change and a former David Lam Chair in Multicultural Education. She is the author of *Hospicing Modernity: Facing Humanity's Wrongs and the Implications for Social Activism* and is one of the cofounders of the Gesturing Towards Decolonial Futures Arts/ Research Collective.

Sophy Banks has spent time as an engineer, radical footballer, and therapist. In 2006 she joined the Transition movement, creating local responses to global challenges. She helped start the first "Heart and Soul" group, anchored "inner" work within the movement, and cofounded Transition Training, taking the transition process to communities around the world. Since 2013 Sophy has cofacilitated shared spaces for grief and trains facilitators in "Apprenticing to Grief" workshops. She also teaches Healthy Human Culture, a trauma-informed approach to cultural design.

Steffi Bednarek is a psychotherapist and consultant in climate psychology, living systems theory, and complexity thinking. She has managed national and international projects, headed up large mental health services, and worked on sociopolitical change for local and national governments, the sustainability sector, and nongovernmental organizations. She is an associate of the Climate Psychology Alliance, a Firekeeper at WorldEthicForum, and an associate of the American Psychological Association's Climate Change Group.

Hāweatea Holly Bryson is a Māori healing practitioner, psychotherapist, Nature-based therapist, and rite of passage guide. Committed to the resurgence of rites of passage, she co-leads global facilitator training in rites of passage and ecotherapy. She is the founder of Nature Knows, specializing in trauma, transition, and transformation.

Bec Davison has worked in the social care and health sector for thirty years with people who are homeless, use drugs, and have been excluded from society because of trauma, poverty, and a lack of opportunity. She currently works in research for the national charity Change Grow Live, actively seeking out the seldom-heard voices, challenging orthodoxy, and questioning traditional research methods to serve communities better.

Inna Didkovska is the director of Kyiv Gestalt University in Ukraine. She is also a psychotherapist with over twenty-five years of experience (Gestalt, psychodrama, and process work) as well as a lecturer, Gestalt teacher, supervisor, author, and organizer of online educational projects including international conferences, international courses, and webinars. She is an active member of EAGT, EAP, and FORGE.

Sally Gillespie is an active member of Psychology for a Safe Climate and the Climate Psychology Alliance, facilitating workshops and writing. Her book *Climate Crisis and Consciousness: Reimagining Our World and Ourselves* explores the psychological challenges and developmental processes of climate engagement. She lives on the unceded lands of the Gadigal and Wangal people in Sydney, Australia.

Wendy Greenspun is on the steering committee of the Climate Psychology Alliance—North America and on faculty at the Manhattan Institute for Psychoanalysis. She has presented workshops and courses for clinicians on ways to work with eco-anxiety and grief. She also provides workshops on building emotional resilience and has run groups (Climate Cafés) for processing climate distress. She has a clinical practice in New York City.

Trudi Macagnino is an integrative psychotherapist and supervisor working in private practice in Devon, UK. She is also a regional academic for the Open University and is completing a PhD with the University of the West of England Bristol.

Julian Manley is a professor of social innovation at the University of Central Lancashire. He has authored and edited books and articles about social dreaming, including *Social Dreaming: Associative Thinking and Intensities of Affect* and (with Susan Long) *Social Dreaming: Philosophy, Research, Theory and Practice*. He is a director of the Centre for Social Dreaming.

Shelot Masithi is a psychology student and young environmental activist from South Africa. She has a passion for collective interventions to address mankind's collective trauma. She is the founder of She4Earth, an organization that is educating young people about environmental crises with solutions rooted in Ubuntu. She is a volunteer at Force of Nature, a social change ambassador at Thred Media, and a YOUNGA 2021 youth delegate. She is also an author and a passionate hiker.

Rebecca Nestor is a climate activist, facilitator, and organizational consultant who supports people and organizations in dealing with the emotional impacts of the climate crisis. Her doctoral work examined emotional experience in organizations that engage the public on climate change. She is a board member of the Climate Psychology Alliance.

Chief Ninawa Inu Huni Kui is a hereditary chief of the Huni Kui Indigenous people of the Amazon and the elected president of the Huni Kui Federation of the state of Acre. He is the dean of the Faculty of Living Systems of the Huni Kui University of the Forest in the Amazon and a research partner in several projects related to Indigenous and Earth rights. He was the International Indigenous Scholar at the Peter Wall Institute for Advanced Studies in 2022 and 2023 at the University of British Columbia. Chief Ninawa is a global advocate against the financialization of Nature and false climate solutions, and the lead author of the ClimateFRAUD framework.

Peter Philippson is a Gestalt psychotherapist, trainer, teaching and supervising member of the GPTI Institute, founder member of the Manchester Gestalt Centre, member of the New York Institute for Gestalt Therapy, senior trainer for GITA, advisory board member for the Centre for Somatic Studies, and a guest trainer for many training programs internationally. He is also a past president of the Association for the Advancement of Gestalt Therapy and the founder of IG-FEST. He is the author of *Self in Relation, The Emergent Self,* and *Gestalt Therapy: Roots and Branches.*

Rhys Price-Robertson is a Gestalt therapist in private practice on Dja Dja Wurrung land in Central Victoria, Australia. He is on the teaching faculty of Gestalt Therapy Australia, and he has published widely on topics such as psychotherapy, social theory, mental health, fatherhood, and family life.

Rosemary Randall is a psychotherapist with a long history of involvement in the climate movement, and she has written and published widely on the psychology of climate change. She is cofounder of the Carbon Conversations project and a founding member of the Climate Psychology Alliance, and she is currently active with Cambridge Climate Therapists.

Chris Robertson has been a psychotherapist since 1978. He is a former chair of the Climate Psychology Alliance and a cofounder of the Re-Vision psychotherapy training organization. He is coauthor of *Climate Psychology: A Matter of Life and Death* and contributed the chapter "Culture Crisis: A Loss of Soul" in *Depth Psychology and Climate Change.*

Gillian Ruch is a Climate Psychology Alliance member who has worked alongside Rebecca Nestor for the past two years engaging in and supporting the development of Climate Cafés. With a background in social work practice, education, and research, Gillian is committed to helping people integrate the psychological and social experiences that the climate crisis is provoking.

Mary-Jayne Rust is a British art therapist, Jungian analyst, and ecotherapist. She teaches, lectures, and facilitates workshops on ecopsychology in a wide range of settings, weaving together the ecological, psychological, political, and spiritual. Feminist psychotherapy helped to broaden her understanding of how culture shapes our internal worlds. Mary-Jayne is author of *Towards an Ecopsychotherapy.*

Nontokozo Charity Sabic is an advocate for climate and social justice, community living, and North–South healing and reconciliation, utilizing the principles of Ubuntu. She works with international environmental and social movements to develop ways of dismantling systems of oppression, decolonizing, and healing internalized racism.

Harriet Sams is an ecotherapist, mentor, animist guide, teacher, and writer. She is a member of the board of the charity Radical Joy for Hard Times. Harriet currently runs "attending to place" workshops through the Tariki Trust, among other ecospiritual teaching modules. Harriet co-hosts Through the Door workshops for the Climate Psychology Alliance. She runs ecospirituality workshops, one-to-one offerings, and ritual circles online and in Cumbria, UK.

Mark Skelding is a psychotherapist in private practice in British Columbia, Canada. He has trained in psychosynthesis psychotherapy and has studied social ecology. He was a faculty member of the Institute of Psychosynthesis New Zealand, and he has initiated experiential ecopsychological courses. He is a community research associate at Auckland University of Technology in Aotearoa New Zealand.

Sharon Stein is an associate professor in the Department of Education at the University of British Columbia. Her research asks how (higher) education can prepare people to respond to "wicked" social and ecological problems in responsible and reparative ways and can support people to unlearn harmful and unsustainable habits of knowing and being. She is author of *Unsettling the University: Confronting the Colonial Foundations of US Higher Education*.

Rene Suša is a researcher-in-residence at the Pacific Institute for Climate Solutions, working in the field of climate education. His research and educational work focuses on the methodology of social cartography as an approach to (collective) inquiry that can make visible the complexity of different positions and understandings related to the wicked and super-wicked global challenges of our time. Rene is also part of the Gesturing Towards Decolonial Futures collective.

Steve Thorp weaves his way through a number of life strands and passions. His writing is published by Raw Mixture Publishing. He works as a school counselor and integrative psychotherapist. He founded *Unpsychology* magazine in 2004 and continues to find joy in editing and crafting it. He cycles, swims, and surfs.

Keith Tudor is a professor of psychotherapy at Auckland University of Technology, Aotearoa New Zealand. His first training was in Gestalt therapy, following which he trained in transactional analysis and person-centered psychology. He is the coauthor of two forthcoming books in this field: with David Key, *Ecotherapy: A Field Guide,* and with Bernie Neville, *Eco-centred Therapy: Revisioning the Person-Centred Approach for a Living World.*

Malika Virah-Sawmy works as a facilitator for systems change with companies, institutions such as the UN, and social and Indigenous movements. She is dedicated to social justice, community living, Nature connection, and North–South healing and reconciliation. She is also engaged in research and storytelling in similar fields.

Sally Weintrobe is a fellow of the British Psychoanalytical Society, a long-standing member of the Climate Psychology Alliance, and the chair of the International Psychoanalytical Association's Climate Committee. Sally edited *Engaging with Climate Change: Psychoanalytic and Interdisciplinary Perspectives*, authored *Psychological Roots of the Climate Crisis: Neoliberal Exceptionalism and the Culture of Uncare,* and coauthored *Climate Psychology: A Matter of Life and Death.*

Francis Weller is an American psychotherapist, writer, and Soul activist. He is the founder of Wisdom Bridge, an educational project that synthesizes psychology, anthropology, and mythology. He has taught at Sonoma State University and the Sophia Center, and he has been the featured teacher at the Minnesota Men's Conference. He is author of *The Wild Edge of Sorrow, The Threshold between Loss and Revelation, In the Absence of the Ordinary: Essays in a Time of Uncertainty,* and *A Trail on the Ground: The Geography of Soul.*

Introduction

Steffi Bednarek

Regenerative Disturbance

Globally, we have entered the ground of disruption, where present and future are impacted by the bleak reality of climate change, biodiversity loss, and the risk of social collapse. The dominant Western culture, built with so much technical genius, seems unable to halt a trajectory into a human-made catastrophe. The hyperindividualized outlook of Modernity is confronted with global problems that cannot be solved from the vantage point of the paradigm that created them. Whoever we are, whatever we do, we are all part of a system that will undergo extreme change in the next decades, one way or another.

This book explores how Western assumptions of what it means to be human may inadvertently collude with a paradigm that erodes the fabric of life. How might these narratives need to be widened in order for us to remember how to participate in the interconnected family of things?

Even though this book is written with a context-specific focus on psychotherapy and the mental health professions, a systems view suggests that complex systems are nested systems. The same systemic assumptions that are present in the mental health field permeate through most professional fields. Largely invisible and unnamed assumptions act like an adhesive that molds everything into the same familiar patterns over and over again. Therefore, before we can engage in meaningful change we need to understand how our basic assumptions may unintentionally hold things in a stuck place. While this is true for individual professional fields, the exploration also needs to transcend the fragmented compartmentalizations of the territory of life so we can figure out together what is asked of us as people and as professionals at this consequential moment in time.

The philosopher Zygmunt Bauman (2000) writes about Western culture as being in a state of "liquid Modernity." Interlocking crises create chaotic, ungovernable situations, where change in one area of the system has unpredictable ramifications throughout. The complexity is staggering, but Modernity is not well equipped to know how to stay with the vulnerability and uncertainty of the situation. Modernity likes narratives of mastery, control, and upward movement and is less comfortable with uncertain trajectories and downward trends. Its way of dealing with complex problems is to break them into multiple individual parts that can be defined, analyzed, measured, and channeled into predictable outcomes. But the solutions implemented in one area quickly become a problem elsewhere. The patterns we try to change keep being built into the change.

This raises the question of how we can think ourselves out of the problem when the way we think is part of the problem. How do we move away from the normative convergence around these self-perpetuating feedback loops and gesture toward regenerative ways of relating to the world and ourselves?

Responses that match the magnitude of the crisis require approaches that are complex, interrelated, systemic, and decolonial. Systems theory tells us that complex systems cannot be willfully directed in linear ways; they can only be disturbed. When the current arrangements are no longer working, derangement is needed so rigid structures can be dissolved and things can have a chance to come out rearranged. The inevitable uncertainty that ensues forces us to slow down, look around with fresh eyes, and reorient ourselves.

As such, this book will not give linear directives or quick-fix solutions. It is an invitation to bring regenerative disturbance to the commons of our profession at the most critical time for our entire species. It is an invitation to slow down, widen the lens, and take a closer look at the ways in which aspects of our professional assumptions and perceptions may serve as the glue that holds stuck patterns in their familiar place.

The heroic stories of eternal progress, human supremacy, white supremacy, male supremacy, Western supremacy, and intellectual supremacy have torn wounds into the fabric of the world that are so deep that they cannot be healed by the system that created them. In recent years we have seen some of the monuments to the ideologies of supremacy tumble and fall. But we also have to ask what monuments have been erected in our minds during all these years of socialization into a fragmented, colonial, and industrialized worldview. What damage has been caused by the legacies of Modernity, and how do they show up in us, our relationships, our ways of thinking, and our professions?

Western scientific traditions have delineated the boundaries of what is deemed acceptable knowledge and have rejected perspectives and perceptions that did not align with the dominant ideology. Good mental health, for instance, is mostly regarded as the ability to function symptom-free within the capitalist paradigm. Alfred Korzybski's work warns how easy it is to forget that the map is not the territory and that the ontological and epistemological maps we use to describe experience risk leading to tunnel vision if we don't zoom out of them from time to time. The territory we deal with is life, and life is transcontextual, interconnected, and alive. It does not happen in disciplines and categories. Of course we need maps to orient ourselves, but maps need to be updated; and most importantly, we must not mistake the maps for life itself. This involves a critical questioning of the ontological and epistemological underpinnings that we have been educated into.

In a time of crisis, we have the opportunity and maybe the responsibility to readjust our habitual ways of doing and seeing things. Before any change can happen, we need to create the conditions in which the system becomes ready to change. As any gardener knows, unless the conditions are right and the ground is carefully prepared, new growth is not sufficiently supported. A composting and fermentation process needs to break down existing structures and add nourishment to the soil.

This book can therefore be seen as a fermentation agent that supports the necessary decomposition of familiar but outdated assumptions. In a diffractive and richly textured kaleidoscope of perspectives, the authors explore how the wider cultural malaise may be reflected in the psychological professions, and they bring solvent to calcified ideas that tell impoverished stories of who we are as a species, how we relate to the world around us, and how we understand who and what needs healing.

As psychotherapists, we know that in a time of change it is difficult *not* to grasp for the habitual strategies that got us into trouble in the first place. Disturbance and disorientation are not states many people would ordinarily choose. The unknown is often the most unbearable of places, and for many even the certainty of catastrophe is easier to bear than uncertainty. But we also know that disruption can serve as a powerful catalyst. We then speak of a maturational crisis or posttraumatic growth, where the crisis serves as a threshold into an expanded identity. Maturity and wisdom often emerge when we are stretched large by facing the aspects of life that break us open. It is in the falling apart that a new self can emerge.

But there is an important difference between a maturational crisis and the rupture of trauma. A maturational crisis requires individuals, cultures, and the psychological professions themselves to master two opposing things: to allow rigid structures and outdated worldviews to crumble and dissolve, and to provide a strong holding container so there is no collapse into fragmentation, breakdown, or premature death. Without adequate support structures, disruptive experiences become traumatic. They cannot be assimilated and are split off in an attempt to numb the wounded part. The problem is that in an individualistic society, many of the collective frameworks that grew over millennia have been lost. It is no longer clear who or what can provide the collective containment that is needed in a time of collective crisis.

For tens of thousands of years, rituals and ceremonies provided the means by which the community addressed the need for healing and renewed its relationship with place. The urge for community sits deeply in our psychic structure, even though the traditions that have bound people together in community have largely been dismantled by an ideology of individualism. This book illustrates that psychotherapy and the psychologically oriented professions can play an important part in rebuilding communal containers and help people to remember how to hold each other in grief, uncertainty, and fear.

Without containing support structures, we risk entering a collective trauma field at global scale. In a traumatized culture, only a certain aspect of society is free to develop, while parts of the culture remain frozen and fragmented. Only if we look through a trauma-informed lens can we see that the level of dissociation and inaction that we continue to witness may not be a lack of care, but an unconscious defense mechanism against the overwhelm of chronic trauma.

Psychology, psychotherapy, and the mental health professions hold important expertise in relation to change, trauma, patterns of repetition, defense mechanisms, and resistance. Our profession has a crucial role to play in supporting people and the culture at large in navigating this threshold moment. But as we know, we cannot help others if we have not undergone the journey ourselves. So this cultural threshold may also present an evolutionary threshold for the profession.

Over its relatively short existence, psychotherapy has gained extraordinary expertise in supporting individuals in their journey through life. The focus has, however, mostly been on the interiority of the purely human experience, which kept the world, the lichen, the oak, the rivers, the cities, and the office buildings out of what we have been interested in. The hegemonic ideology of separability has isolated us from our sense of belonging to a greater, possibly more meaningful entity. In an individualized psychology, the problem becomes interior, and we ascribe our symptoms to a personal shortcoming that we try and fix or "work on." The cultural obsession with things rising is often mirrored in the idea of perpetual self-improvement. There is an entire industry of self-help books and lifestyle advice that is underpinned by the idea that our lives are meant to be extraordinary and that we always deserve more. We miss the value of the ordinary and risk pathologizing the aspects of life that refuse to move upward or that lead us downward. But what if our primary human need is not to attend meticulously to our individual needs, but rather to live our flawed and imperfect human lives in a participatory way, in continual relationship with all that lies outside ourselves? This would require an unlearning of many ideas that Modernity has taught us and entail a move toward decentering the self in service of community—including the community with rivers, trees, and foxes.

The Greek word *psyche* means Soul, but we rarely mention Soul in the mental health professions anymore. We have become so much more interested in the self. In the opening chapter of this book, Francis Weller states that we have transitioned from a Soul-focused "psyche"-ology to a "self"-ology. And this self is mostly not viewed as an entangled ecology of relationships, but as a separate, coherent, and contained entity. With this shift it is easy to forget that what we call "I" is made up of more nonhuman elements than human cells and that there are other agents out there that shape the world and our own lives.

The decision of whether the focus needs to lie on the individual, the environment, or the complex ecological coevolution of systemic interrelationships is of course not an either/or decision. From an ecological lens, all perspectives are needed and are in dialectic relationship with each other. It is just that some have been favored over others for a very long time. This monoculture eventually creates imbalance and a certain toxicity level in the ground.

The focus on the individual is the in-breath, but the out-breath needs to spill out into the streets, into the education system, urban communities, forests, rivers, and the beauty of the world.

When I talk about "the world," I am not talking about commodifying Nature as a new resource that we can "use" to make us feel better. I am talking about the reciprocity with the world that is invited into the relationship. And so we may need to include shopping centers, multistory car parks, manicured lawns, and refugee camps in our consideration of health, well-being, and the notion of who we are.

James Hillman (1995) says: "By not recognizing that the soul is also in the world, therapy cannot do its job anymore. The buildings are sick, the institutions are sick, the banking system's sick, the schools, the streets— the sickness is out there" (pp. 3–4). The "great turning" (Macy & Brown, 2014, p. 6) may therefore not be a call to flock to Nature but rather to turn our attention toward a meaningful engagement with what lies outside us. It is the capacity to move from the skin boundary to a state of interbeing, cultivating the capacity to fluidly hold both polarities simultaneously: the vast complexity of the human and nonhuman world out there and the experience of this singular point of contact in this unique body.

Systems theory encourages us not to abandon old modes of thought in favor of the novel, but to stay with the chaos of complexity and the discomfort of the unknown for long enough to see what emerges beyond familiar patterns. As humans, we have the capacity to sense into atmospheres, to feel, to imagine, to vision, to smell, taste, dream, and perceive in myriad different ways. These are the faculties with which we contact the world. Our sensual bodies, our feelings, and our imaginations allow us to perceive and transmit information that the rational mind cannot grasp. This countercultural engagement disrupts the cause-and-effect mindset and demonstrates the paradoxical nature of change. After all, it is at the edge of chaos that complex systems are at their most creative. New ways of seeing require a willingness to hear more than the echo of familiar thoughts and an openness to relate to the world in ways that allow the relational space to reveal itself to us.

In the final chapter you will find a range of practices that take the focus off the familiar grooves of the individual and invite a step into relational patterns that open up new ways of seeing. These interventions are examples of responses to an ailing culture, a culture that most of us are part of, that many of us are suffering from, and that most of us participate in.

Before I bring this introduction to a close, I want to point out that the structure of this book has consciously moved away from individual, self-contained contributions. Everything is in conversation and relationship. The main themes weave their way through the entire book, in and out of different forms of expression—from the personal to the collective and from academic thought to poetic imagery. The contributors come from a diversity of contexts, continents, cultural backgrounds, perspectives, and experiences, and yet there are common themes that run through everything. The hope is that some ideas may leave the contraption of the page and cross-fertilize whatever seeks expression through you, the reader, finding their way into the part of life you are in contact with. We hope to encourage you to add your own voice to the evolution of our profession in response to what the current times are calling for, weaving back together what has been torn apart. Let's keep the conversation going.

How Did We Get Here?

In 2021, the *British Gestalt Journal* invited me to guest-edit a special edition on climate change. I started this journey with an invitation to a group conversation and was excited that four psychotherapists whose work is influential in the field of climate psychology—Bayo Akomolafe, Mary-Jayne Rust, Sally Weintrobe, and Francis Weller—accepted straightaway.

Even though most members of the discussion group ordinarily have large public-speaking engagements, this conversation was private. There was no expectation for it to serve anything other than our coming together. We gathered in an incubation space around a single issue: *the role of psychotherapy in a time when the familiar is dying.* Nobody led the conversation; there was no prodding or probing. We sat with this single question and paid attention to what emerged between the silences. Clusters of meaning emerged organically.

After the conversation took place, all participants agreed for the conversation to be transcribed and to form a provocation paper for the special edition on climate change. The transcript of the original discussion is the opening chapter in this book.

The theme had clearly struck a nerve, and the conversation continued beyond the journal's special edition. The publishing house Confer and Karnac offered to expand the inquiry into a book project. Many more colleagues from all over the world joined the discussion by submitting responses to this central question. The book was due to be published in May 2023, but sadly the publisher went into insolvency one month before the publication date. The book itself experienced the collapse of the expected familiar trajectory. Its path illustrates the content.

I am grateful that North Atlantic Books were excited to offer a new publishing contract, calling the book "groundbreaking." I am grateful for their support, encouragement, and enthusiasm. My special thanks also goes out to the *British Gestalt Journal* for their permission to republish all seven articles that were part of the journal's special edition, which was published in November 2022.

My hope is for the discussion to continue. Many of the contributors have offered to enter a dialogue with the readers of this book after it is published. To that end, you can visit the website www.climatepsychol ogyandchange.com to find a program of events, talks, and experiential workshops that allow cross-fertilization and a further engagement with the book's central themes.

Overview

This book is divided into seven chapters, all of which explore the main title question from different angles. It is important, however, to point out that the major themes of individualism, trauma, anthropocentrism, fragmentation, and decolonial processes run through the entire book. They intermingle and weave themselves in and out of different chapters. The division into chapters is therefore almost arbitrary, giving the mind some form of structure while also allowing permeability.

The book opens with the transcript of the original discussion between Bayo Akomolafe, Mary-Jayne Rust, Sally Weintrobe, Francis Weller, and me (Steffi Bednarek) on the role of psychotherapy in a time when the familiar is dying. The conversation covers much ground, but the main themes range from the ideologies that have been passed on in psychotherapy training to an exploration of the dialogic relationship between individual and collective trauma, the long shadow of colonialism within the profession, the evolution of therapy, and the need to step out of the consulting room and into the world.

Chapter 1 focuses on the shift from the individual to the collective lens. Steve Thorp asks whether things have to be the way they are within the profession of psychotherapy and how they would be if we were not bound by the paradigm we are troubled in. He points out that any therapist who seeks to step beyond the way things are faces the dilemma that professional validation, accreditation, and development have a tight grip on what lies within the accepted boundaries and codes of practice. Thorp proposes that healing not only is found in the spot that hurts but also occurs in the space between straight lines.

Chris Robertson moves the psychological focus from the individual to the culture and from personal growth to decentering practices and kinship. He proposes that individual symptoms are the openings that lead us to a larger story. We need to listen through them, in an attempt to hear the unspoken, unseen, and uncontained pathos of an ailing culture. Robertson draws our attention to the need for collective rites of passage and the shift of mindset from the Anthropocene to the Humilocene.

Chapter 2 invites the state of the world into the consulting room. Trudi Macagnino's contribution explores the social construction of silence around the climate crisis in the therapeutic encounter. Her research findings show that both therapist and client are frequently defended against overwhelming feelings and that therapists need to have worked through their own anxieties in order to support clients in facing the state of the world. Wendy Greenspun offers a fascinating case study that provides insight into her work with a client who suffers from climate distress. As Macagnino's research shows, it would have been easy for the therapist to indirectly shift the focus to a more comfortable terrain, but Greenspun movingly discusses her own journey of "staying with the trouble" and her willingness to allow herself to be exposed to the existential threat that affects her as much as her client. Both authors draw on their psychological expertise while also openly sharing their own sense of lostness and distress.

The causes and impacts of the climate crisis are distributed differently depending on where we live and where we sit in the capitalist and colonial power structure. **Chapter 3** specifically focuses on the long shadow of colonialism in the profession of psychotherapy and illustrates how climate change and colonialism are intricately related. Reports about the negative effects of climate change almost exclusively focus on the impact on the global North, placing more value on white lives and culture than on Black and brown lives and culture. Those of us who live in the global North have the luxury of being worried about our children's and grandchildren's future, while climate change has already affected vast amounts of people in the global South. The freedom to be protected from suffering in the Western world is based on the unfreedom of people across the rest of the world.

The contribution of African climate activist Shelot Masithi highlights that the concern about future traumas is a concern of the privileged. Masithi's chapter describes her early experience of living with water scarcity in South Africa. She describes the importance of Ubuntu in bringing us together and ensuring the survival and well-being of everyone. Ubuntu is a reminder that a person becomes a person through the community of beings and that nobody gets to be a moral being separate from their community. Ubuntu also connects Masithi's writing with the contribution from Nontokozo Sabic and Malika Virah-Sawmy, who refer to the South African philosophy of Ubuntu in relation to their decolonizing work in the environmental movement. They stress the importance of recognizing that white supremacy affects all of us. Virah-Sawmy recounts how long it has taken her to be able to see and name the visibly invisible shadow of colonialism within the climate change movement.

Hāweatea Holly Bryson raises awareness of the fact that Western notions of healing take little notice of the wisdom of other cultures. She stresses that decolonizing psychotherapy requires more than ensuring equal access to opportunities within a Western paradigm. It requires a dismantling of the colonial power structures that are inherent in the profession itself. It also requires an acknowledgment of the unacknowledged shadows in the lineage of our profession and a widening of the notions of healing, so that the wisdom traditions that precede psychology and psychotherapy are no longer exiled, denigrated, and eradicated. Bryson's chapter is a call for the profession to create respectful space for other cosmologies, ontologies, and epistemologies.

The healing of wounded psyches and the restoration of damaged landscapes is a global imperative. This healing journey can no longer maintain the split between the human and the more-than-human world; instead it needs to consider that neither can heal without the other. **Chapter 4** is a reminder of the urgent task to unlearn anthropocentrism and to decenter the human. Matthew Adams takes the discussion into a radical relationality and animist epistemology that expands the notion of healing into the realm of reciprocity, cooperation, and interdependence between humans and other-than-human beings. He draws on research in psychology and related fields to explore ways in which human health is entangled with other-than-humans. Rhys Price-Robertson, Mark Skelding, and Keith Tudor also draw our attention to the entanglements of psychotherapy with the anthropocentric worldview. They demonstrate how psychotherapy can facilitate the transition from a human-centered perspective to an ecocentric one, and they offer reflections on the clinical implications of an expanded notion of self that sees humans as one member of a vast family of beings.

Glenn Albrecht proposes a direction of travel from the dissociation in the Anthropocene to association in the Symbiocene. He describes a process that he calls *sumbiography,* which documents the intersection of personhood, other people, place, and nonhuman life, which reveals either the richness or the impoverishment of these interrelationships. A sumbiography can reveal unresolved trauma with respect to Nature or an absence of contact with Nature and wild things. The use of a sumbiography can help a person move from a state of dissociation from Nature to one of greater association.

Chapter 5 addresses collective trauma, fragmentation, and the risk of social collapse. I (Steffi Bednarek) use the perspectives of brain hemisphere balance, collective trauma, and the concept of necessary suffering to explore the personal and cultural fragmentation and compartmentalization that can lead to inertia and inaction. I suggest that pockets of Western culture are built on layers of unprocessed collective trauma and that the divided and polarized stories that dominate Western culture serve as a defense against overwhelm. This dissociation allows ordinary human beings, not monsters, to carry out monstrous acts, and it allows the horrific to become normalized through small steps into the unthinkable. This chapter identifies a need for collective ways to reclaim fragmented parts, bridge existing polarities, and meet challenges with the kind of maturity that is gained by accepting necessary suffering.

We witness that the global metacrisis includes increased polarization, a destabilization of democracy, and a rise of authoritarianism on the world stage. The risk of social and political instability presents a very real danger. How do we prepare for a situation when we can't know what we need to be prepared for? And how do we increase resilience when the familiar pillars of support are in a state of collapse? What Ukrainian therapist would have thought before February 2022 that they would soon need to find ways to survive in a war while also supporting clients who live in a war zone? Through the experience of the war in Ukraine, the following two articles explore what support is left when environmental support vanishes.

Peter Philippson was meant to give a series of webinars for Kyiv Gestalt University when Russia invaded Ukraine. He was asked to go ahead with the webinars and unexpectedly found himself addressing more than six hundred practicing psychotherapists who had now also become soldiers, refugees, or volunteers in a war zone. His chapter is an edited transcript of his four lectures, which cover the themes of aggression, creativity, resilience, self-support, trauma, hate, and living in the conscious presence of death.

Inna Didkovska, the director of Kyiv Gestalt University, then shares her direct experience of being exposed to the trauma of war herself while also teaching, supervising, and counseling clients and trainees who experience war. Her chapter is a reflection on the existential nature of meaning-making. She asks which means of support are left in field conditions that strip people of their usual support structure. Inna's writing highlights that there is power and "response-ability" in facing difficult truths.

Chapter 6 illustrates how facing difficult truths in relation to climate change can alter and expand psychotherapeutic practice if defenses are lowered and therapists allow themselves to be affected and undone by the seriousness of the threat. Once the reality sinks in, business as usual is no longer an option, and the compartmentalization between work and the rest of life no longer makes sense. This chapter tells the personal story of two senior climate psychologists, Rosemary Randall and Sally Gillespie, and their journey of creative adjustment to a dramatically changing world.

The COVID-19 pandemic changed the practice of psychotherapy for good. Without time to prepare, the profession was forced to step over a previously unthinkable threshold and allow many of its carefully constructed professional constructs to be disrupted. The sanctity of the therapy room and many traditional professional boundaries were dismantled from one day to the next. Therapists started to work from home, and clients called in from their cars, bathrooms, kitchens, or park benches. Children walked in, washing machines beeped, food deliveries needed to be prioritized. Client and therapist were suddenly affected by the same global event. At times clients were in a more supported position than their therapist. The climate and biodiversity crisis has not yet entered the everyday lives of the majority of people. When it does, the effects will be immeasurably worse than the COVID pandemic and will claim more lives. Responses to the catastrophic consequences need adaptations that go beyond familiar practices. Sometimes the call for change comes in unexpected ways.

Long before the pandemic, Sally Gillespie and Rosemary Randall allowed the horrific reality of climate change to disrupt their professional paths. Their capacity to face unbearable truths became a catalyst for change. Sally Gillespie describes how dreams have guided and encouraged her to step through the door of the consulting room and to bring psychological expertise into the wider community. Rosemary Randall describes her journey of initiating the much-acclaimed "carbon conversations" and explains how this work has transformed into a new regenerative community practice.

Chapter 7 explores community practices that aim to disrupt the linear, cause-and-effect mindset and increase the capacity to hold complexity in all of its wild and paradoxical forms. A powerful range of practical approaches illustrates how we can shift the focus from the individual to the wider culture and how we can diffract habitual perceptions and step into different relational patterns.

Vanessa Andreotti, Rene Suša, Cash Ahenakew, Sharon Stein, and Chief Ninawa Inu Huni Kui describe some of the deeply transformational practices that the Gesturing Towards Decolonial Futures collective takes people through. Their aim is to raise awareness of the complex interplay between the cognitive frameworks and affective imprints that most of us have been socialized into within modern/colonial systems and the ways in which these maps shape responses to complex problems and influence the potential mitigations that present themselves in relation to the climate emergency.

Through a series of interventions, participants gain increased perspective on the ways in which the concepts of Modernity have put an imprint on their self-images, including hopes, fears, insecurities, and investments; their immediate context and perceptions of historical, systemic, and structural forces; their generational cohort; the universalization of the social/cultural/economic parameters of normality; familiar patterns of relationship building and problem solving; the habit of elevating humanity above the rest of Nature; and the urge to find immediate solutions while being aversive to states of uncertainty and complexity.

This is followed by Rebecca Nestor and Gillian Ruch, who describe how Climate Cafés offer people an opportunity to take the very first steps toward breaking the socially constructed silence around climate change. Climate Cafés provide a space where fears and uncertainties about climate change can be expressed and explored in the company of strangers. It is a space to pay attention to one's personal experiences and responses in relation to the climate and biodiversity crisis without needing to understand the science, engage in solutions, or feel or act in any particular way. The facilitator's careful attention to keeping the relational space open and free of expectations frees people up and paradoxically creates the ground for a deepening engagement.

Julian Manley introduces the powerfully potent social dreaming matrix, where participants enter an almost dreamlike space while sharing dream fragments. The dreams no longer "belong" to the dreamer. As the images start to mingle with those from other dreams, new associations emerge. This process is particularly helpful for approaching almost hopelessly complex issues, such as climate change, as it allows people to slip out from behind their individual lens and hold the complexity of a collectively generated image that effortlessly incorporates contradictions and changeability. The dream images become a window that gives a glimpse of the hyperobject. Without being able to explain everything, the dreamers generate a sense of bonded thought and an understanding that can feel cathartic or even spiritual.

This process is not dissimilar to what happens in a Warm Data Lab. "Warm data" is a term coined by Nora Bateson to describe contextual and relational information about the interrelationships within and between complex *systems*. Information does not arise in single contexts but moves between multiple contexts that overlap in living communication. Warm Data Labs are group processes, designed to illustrate the interdependence of systemic patterns for people with no previous exposure to systems theory. Bec Davison and I (Steffi Bednarek) describe this kaleidoscope of conversation in which participants move between small, constantly changing groups while reflecting upon a given question through multiple frameworks, such as the economy, education, health, technology, family, and others. Through ever-increasing complexity, the focus of attention is widened beyond the usual frame of reference, and participants start to hold multiple perspectives at once.

Harriet Sams then turns our attention to a relational connection with place. We often hear that we need to love Nature in order to care for it. But how do we extend care for places that carry the scars of the trauma that has been inflicted on them so deeply that we recoil from them? How do we take care of them, and how is their wounding entangled with our own wounds? Sams introduces the practices of Earth Exchange and "attending to place." In both practices, participants are invited to turn toward wounded places in their environment. When a place is hurt, the people who live in that place are usually hurt too. Harriet illustrates that turning toward wounded places, rather than turning away from them, changes our engagement with self, community, and land as part of a relational web.

In a world scarred by the effects of human activity, the connection to wounded places is likely to give rise to grief. Sophy Banks illustrates how it may look when we attend to this grief in community rather than suppressing it or attending to it individually, as though this grief was privately owned. Banks introduces communal grief-tending practices and illustrates how the expression of grief is influenced by the specific culture a society creates. She describes how coming together to express, witness, and make meaning of pain contributes to the health of a community. When it comes to harm caused to communities or the more-than-human world, it often takes a group context for the (collective) pain to be witnessed and to find expression. Shared spaces therefore provide an essential opportunity that allows the wider impacts of harmful systems to be fully acknowledged and felt. Sophy describes how the intensity of sharing grief within the context of relational connection to others is matched by the intensity of compassion and love that arises when we meet each other in our profound vulnerability. A renewed sense of resilience arises."

This book addresses some of the ways in which our perceptions of what it means to be human may need to be widened and stretched in order to be able to hold an ever-larger aspect of the complex, paradoxical, and interconnected whole in our hearts and minds. Our actions will differ depending on what we are capable of including in our view.

All of us alive today live at an evolutionary threshold. Whether we are headed toward a more sustainable future or moving toward our own demise is yet unclear. Investing in a future we want to live in does not mean that there is an automatic guarantee of success. But one thing is for certain: we all have a choice of whether or not we want to take an active part in creating the conditions that allow life to thrive, regardless of the outcome. Rebecca Solnit (2016, p. 5) reminds us that "the future is dark, with a darkness as much of the womb as the grave."

Opening Dialogue

Psychotherapy in a Time When the Familiar Is Dying

A conversation between Bayo Akomolafe, Steffi Bednarek, Mary-Jayne Rust, Sally Weintrobe, and Francis Weller

Hosted and edited by Steffi Bednarek

What follows is the transcription of a dialogue that took place on January 14, 2022, between five psychotherapists from different continents, modalities, backgrounds, and perspectives. In a deliberate departure from the singular voice, this was an invitation to come together across disciplines and to cogenerate entangled and interconnected meaning. The primary aim was a cross-fertilization, a building of bridges across modalities, allowing multiple perspectives to coexist in a rich and textured conversation.

The five participants gathered in an incubation space around a single issue: *the role of psychotherapy in a time when the familiar is dying.* Nobody led the conversation. The participants sat with this single issue and paid attention to what emerged between the silences.

Clusters of meaning emerged from the conversation. They formed a provocation for a call for further contributions. This was the starting point for this book.

The edited video recording of the original discussion is available at www.climatepsychologyandchange.com.

The Conversation

SB: Thank you for accepting this invitation to join me in an incubation space around the central question of the role of psychotherapy in a time when the familiar is dying.

SW: I just want to thank you, Steffi, for getting us together. It's a very nice invitation, and it sets up some expectation and hope of a real intermingling.

BA: I'm also grateful to be here. I feel like starting with what grounds me these days, and that is my family, my people, the community here in Chennai [India], and the promise of doing something entirely incomprehensible in these times. So, it's good to be here with you, my odd kin.

FW: I've spent the last forty years trying to see what trails Soul is laying down in our individual, communal, cultural, and planetary lives. Seeing if I can catch a scent of what Soul is trying to say to us in these times about what we have forgotten, what we've neglected, and what we need to tend. It is good to be here with you.

SB: This morning I went for a walk in a linear mindset. When I looked up, the sheer beauty around me was so breathtaking that it jolted me out of my thoughts. Sometimes I find it difficult to open myself to beauty. How can I contain so much beauty? Who and what can I hold onto in order to also feel the beauty in this time of uncertainty and trouble? This question is very much with me.

MJR: So much is with me when you talk about those things, Steffi. When I really woke up in my body to the ecological crisis, I felt very passionately about our relationship with the other-than-human world. Being immersed in the field of ecopsychology, I was running courses on the west coast of Scotland with an outdoor educator. One of the things that was central was how easily people could recover their deep love for the other-than-human world.

What is with me is a need to rethink child development, which is so much framed in terms of human relationships. The whole of the psychotherapy world is. How do we reframe that in terms of including our relationship with the other-than-human world and how that sits in our inner worlds, too?

SW: My journey was profoundly influenced by Rachel Carson's *Silent Spring*. It shocked me out of myself and put me in touch with my deeper feelings for the nonhuman world. I also began scanning for the presence in the landscape of these new megafarms that so ill-treat animals. I would be looking out for large warehouses in the countryside and wondering what was inside them. That has never left me.

I have grandchildren. One of them, aged seven, said to me recently that there isn't a word that can describe humans and other animals together, so she's going to say "rabbit-person," "cow-person," and "human-person." Children are fabulously sophisticated. Not only are they not so damaged in their relationship to the nonhuman world, but they are also not so racially fractured or class fractured. They are already quite savvy about the way their parents are caught up in the dominant culture. Those are strands that I've carried with me.

Psychotherapy Training and Dominant Assumptions Passed Down through Training Institutes

MJR: I would say that there was nothing in my training at all about our relationship with the other-than-human world.

BA: The university system at the Neuropsychiatric Hospital in eastern Nigeria, where I received a BSc, an MSc, and a PhD in psychology and trained as a psychotherapist, was connected with the church. I didn't need a degree to notice that there was something irresistibly theological about the practice of therapy that situated and reinforced the gilded interiority of the human subject. Everything else was easily dismissible, discardable, pathologizable, invisibilized. And I was scared to go to the places of African traditional systems that were now rebranded as "demonic."

For my research, I went with my *DSM* brain to the shamans and the priests, and I listened to them as they turned my thinking upside down. They asked: "Why would you think of what I would term 'auditory hallucination' as something to be fixed? What if that's your grandmother or your grandfather? What if that's a plant trying to speak with you?" This was really jarring for me. It was disturbing.

Steffi, you spoke about beauty when you started to speak, and this idea of *engorgement* comes to mind, where the membrane can no longer take the weight of its content. In the same light, I think the disciplinarity of my training and my degrees and the things that I was educated to think of as "reality" are now hollowing out. And in their place I'm listening to tortoise and rock and grandmothers and shamans and a very conspiratorial notion of *affect:* that feelings are not domiciled in easily convenient consumerist boxes inside Mary-Jayne or inside Francis, but that affect is *environmental;* that to *feel* is to be enlisted in territorial movements. This changes the context, and now I am in the face of the animal or the inscrutably invisible. That is where I feel a different response is warranted. Worship, maybe.

SW: I've practiced as a psychoanalyst for decades. I certainly think that there has been a great underattention paid to external reality. I don't think it's necessarily in the bones of psychoanalysis and its theory that it needs to be like that. After all, it's a theory that is meant to be addressing external and internal reality and how they shape each other, but it's as if it got taken over.

While my training did not pay much attention to the external world, I was fortunate enough that my analyst was touched by the beauty of the world, which was a blessing to me.

My struggle is that I've got a great deal of value from my training, and so how do I hold those tensions? One needs a discipline and a base of study. One needs not just to be wild about what one thinks. What my training gave me is a way of looking at things long enough so that you suddenly realize how much you don't know, and you get lost, and you really study. We have so much theoretical work to do. We need to learn how to apply that in different contexts, and that's what I've been trying to do in my work.

I would say that trainees need to ask their institutions to address these issues as part of their training, and make a fuss about it. It's really important. All of us must try to get our institutions to address these issues, responding to the urgency of our time.

FW: I was licensed way too young. At twenty-seven years old, I was given permission to sit with people and work with their psyches. But I was smart enough to know I didn't know anything. So I found this "old" man; he must have been sixty. His name was Clark Berry, a Jungian psychotherapist.

The first time I sat down with Clark, he reached over and he patted this big rock that he had by his chair. He said, "This is my clock, I operate at a geologic speed, and if you're going to work with the Soul, you need to learn this rhythm, because this is how the Soul moves." Then he pointed to his wall clock and said, "It hates this." That was the single most important piece of training about sitting with people: how do you teach them to slow down?

BA: What you say about slowing down is so very vital to me that I want to dance on the streets. I often say: "The times are urgent, let us slow down!" Slowing down, for me, is a reframing of our crisis, a reframing of the climate chaos, of the pandemic, even of racial matters. It's a reframe of continuity.

When people hear me say that, they imagine that I'm asking them to take a break, to do more yoga, or to go on a vacation. That's not the ontology that I speak from. The ontology that inspires that saying is one of crossroads.

The Yoruba people in western Nigeria traveled across the Atlantic and infected those slaveholding communities and became the Afro-Asian vibrant communities that we all know and love today. They speak about *orita. Orita* is the crossroads. The crossroads is the place where bodies are refused their coherence. Basically, in an entangling and entangled world, the self cannot be coherent, the self is always diasporic, is always spread out, is always traveling, always emerging, always becoming. To situate oneself at the crossroads is to refuse the salvific impulses of our discipline; the efforts to save, which is often a colonial attempt to reinforce continuity, to reinforce the "normal."

I feel that this is the time for the incomprehensible, and this is sprouting in ways we don't have language for. So the crossroads is an invitation to notice the others in the room. That is the reframe that connects climate change, ecopsychology, feminist insights, this pandemic and its epidemiological reconfigurations. It just turns it on its head and invites us to notice.

We're now in alien territory, and we must now frame our politics and our therapeutic gestures in a way that brings us to these *others,* these more-than-human *others* that have always been the condition for our thriving. What might that look like? That is the question that I'm sitting with.

Individual and Collective Trauma Inherent in the Field

FW: I was sitting in a class for one of my CU units and it was all about these internal practices to regulate trauma. That is wonderful stuff, but as I was sitting there, I was thinking about the thousands of years where it was the community and Nature that were the regulating mechanisms to address issues of trauma, injury, and wound. Ritual was the primary means by which we sutured the tears in the psyche. So maybe we need both. I'm not saying either/or; I'm just saying, what would happen if psychology began to remember its roots?

SB: I wonder whether there's a third that needs to emerge, between looking back and staying within the existing lens. Jurgen Kremer talks about the fact that we all have roots somewhere, but for white Europeans the route back to these roots needs to go through trauma, through the trouble that our ancestors have caused and experienced. There is no way that doesn't lead through this still largely unprocessed collective trauma.

We may meet a time where something new is asked for that lies in the field between the polarities we operate within. Between the polarity of individual and collective, there may be an emergent third. We may need to remember, attend to what we know, look at the bigger picture—and maybe nothing is quite enough, nothing quite fits.

Sally, you talked about your granddaughter, saying we haven't got a word for the in-between, the connection between humans and animals. Maybe it is the unexplored spaces between what we know that need more attention.

MJR: This reminds me of a major trauma, which was about my dog whom I had known from when I was three. She very sadly drowned in the well in our garden when I was twelve. It was hugely traumatic for me, as you can imagine. I never mentioned it in any of my years of analysis. It just got completely overlooked. During all my years of therapy, I told my story of human relationships, but I hadn't told my Earth story. So I felt I had to go back and expand my story about relationships and include the rest of life.

SB: That seems so important. I was born in East Germany, and part of my personal story is hugely political. There were political prisoners in my family, people were spied on very openly, and the whole political system put its imprint on me. In my early twenties, I worked for the Council of Europe on Human Rights, and I worked in a very political way. Then, in an attempt to understand more about human behavior, I studied psychotherapy, particularly Gestalt psychotherapy. Gestalt has the notion that self is not static . . .

BA: That self is *ecstatic!*

SB: Ecstatic, yes! And self forms in constant exchange with everything that is around us, in exchange with the field. This was the theory that I loved, but the practice had almost an invisible boundary around it, where nevertheless my focus was very much on the individual. So the way I told my story also became very interior. It didn't become the story of being human, it didn't become the story of being German, it didn't even become the story of my family system. It became *my* story, and in a way I feel like there's an additional layer of trauma that was added because of this isolation, the fact that I became so myopic.

The widening out into collective trauma that is not privately owned, that I may find out more about *me* when I look at others or when I look at the culture, is so important. Even in trauma, there's something deeply moving and joyful when it becomes larger than my own story. It makes it much easier to contain things.

And in terms of climate change, that's my experience, too. If I just think about it in terms of what's going to happen to me and my loved ones, then it's a very isolated story. Widening this out, there's beauty that is possible.

BA: Beautiful, Steffi, thank you.

FW: I've noticed over the past few years how much the material that comes in the room is changing. It's not so intrapsychic anymore. It's not so much talking about my history, my wounds, my traumas. The world is coming into the room. Clients are talking about economics, the climate catastrophe, and their impinging fears. The ambient field of anxiety, uncertainty, and grief is really coming into their psychic lives.

The patient is now the community, the patient is now the culture, the patient is the planet. It is not so much my individual personal wounding any longer. That's in there, obviously, but there's a sensitivity coming through that crease in the Soul.

What's really coming out of the people I'm sitting with is the larger ambient field that the therapist themself is not immune to. I mean, who is the patient in the room now? I'm as anxious as anybody I'm sitting with. I'm as grief-stricken as anybody I'm sitting with.

We cannot touch this stuff without the presence of community. How do you hold this grief by yourself? It's insane! We've been isolated in our grief, isolated in our longings, isolated in every piece of our existence. That itself is the pathology we have to overcome, how to make this all communal once again. And still, there is no answer. But at least there's five of us on the screen right now. I don't feel alone in this moment with my tears, with my grief, with my fear. That's the solitary confinement that comes out of the Western ideology of individualism. You're basically on your own.

SB: James Hillman said something along the lines that psychotherapy is stuck because we don't see that the institutions are sick, that the banking system is sick, the education system is sick, that the sickness is out there. It's not doing away with individual psychotherapy, but I don't know the tools for working with the sickness in the culture.

MJR: How do we enable the personal to resonate with what's happening with the collective? It's not that the focus has changed the collective, but how do we build bridges between?

I spent ten years working at the Women's Therapy Centre, looking at wounds around eating. We live within a patriarchal society. Our problems come out of our family history and the culture that we live in. One way of framing what we're living through is that in terms of consumer culture, we're living in a giant eating problem, all together. How do we work our way through that? What does it mean to attend to our hunger? What are we really hungry for? There are all those questions in there, for me.

SW: Yes, and in order to hear what is traumatic, the therapist has to take the trauma into themselves and work it through themselves. If one is with a therapist who is closed to experiencing trauma, the patient is in trouble. Patients who have been traumatized know whether they can raise something. If they're not going to be heard, they will be traumatized further, and they know that.

I don't call myself a recovering psychologist as you do, Bayo, because that implies that it's something that I see as only dangerous. While I absolutely go along with the truth in that, I also find myself undone with questions like: How do we keep a perspective on the individual *and* make more allowance for how feelings exist beyond the individual? How do we not collapse things? How do we not get fractious about this in nonproductive ways? I think we're at a point where a great deal of theoretical advance can be made in our profession if we hold these tensions.

BA: Sally, I deeply value the invitation to hold tensions and diffraction instead of thinking in terms of binaries, which is problematic. You want to see how things come together and fall apart together, places of convergence and divergence.

I think the rhetorical force of speaking with the cadence of the trickster and saying that I'm a recovering psychologist is commensurate to the political needs of my context. The world that I come from skews heavily toward the colonial and the imperial and is still battling with feelings of shame.

We have lost rituals, celebrations, and patterns of communing and listening to each other. We curse each other on airplanes flying into our airports, and we say things like: "Why don't we look more like New York?" or "Why don't we look more like London?"

The invitation that I hold is to say: "No, we don't need to subscribe to an image that has been imposed on us. Let's compost that image and see what sprouts from that soil!"

SW: Thank you for that. So many of us are in recovery from that kind of self-loathing, which is a cultural product. Thank you, that's really clarifying.

The Long Shadow of Colonialism in the Psychotherapy Profession

SB: I am aware that monuments to colonialism have recently been toppled in many places, but there's an ongoing question about how these monuments have been erected in me, in my thinking, in my mind. And how do I participate in keeping them alive? What needs to crumble in me? And what facilitates that process? Some of this is hardly tangible.

SW: That whole move—and Steffi, you mentioned that it's linked with private practice—is linked with privilege. It's linked with the fact that in the global North, therapists are largely white. It's linked with a whole culture that has been very splitting into the haves and the have-nots, the better and the worse and all the rest of it. A fracturing culture.

I don't think that the analytic and therapeutic profession has been any different from general culture in that sense. The patient and analyst can be caught up in a psychic retreat from reality, very much in the service of maintaining privilege within a culture.

FW: At the heart of all of our sorrows is this profound sense of emptiness. This is part of the legacy of *white amnesia* and the forgetting of culture, and the trauma of rupture between cultures. And how much we've been cast out of a sense of continuity and connectivity to place, to language, to traditions, to practices and rituals. We've been left in that legacy of individualism, basically with an empty self. And so much of our attempt to cope with that emptiness has been consumption. Not just of material goods but also of power and racism and everything we can do to somehow avoid the confrontation with that emptiness.

I think, as therapists, we also have to examine our own experience of that in our traditions and in our field, so that we're not also contributing to it. How do we navigate around the emptiness?

Cultural Entanglement and Private Practice

MJR: How can I bring all of these different threads into my work as a psychotherapist? I don't believe in introducing things deliberately into the work, so how can I listen for it? How are my clients bringing it?

FW: Because when clients come to us, there's an urgency to change, but that change is almost always predicated on self-hatred: "I'm not good enough, I have to change who I am to fit into the construct of approvability." Because we have forgotten our collective sense of belonging and we have become so isolated into our singularities.

We're anxious all the time, whether "I'm in" or "I'm out," and people come to therapy so much to see if they can craft a self that fits the expectations of belonging. So, a lot of my work with clients is trying to disavow that fantasy, that fiction of "how do I fit in?"

SW: I think the job of therapist has never been more complicated or onerous. What we have to be burdened by has never been more difficult. It undoes us, but I still think, if we open ourselves up to this, we will have fantastic theoretical advances, which are needed. All this is holding back the profession, and we will also be of more help to the real problems that people are facing now.

BA: There's hardly any choice we have left but to slow down and to build a classroom around the cracks. And maybe the new will happen to us. Not just depending on our own skills or genius to weave it into being.

My people say that sometimes wisdom is speaking out of the corners of one's mouth. There's something about speaking that always ruptures something else. There is no middling area, there's no Socratic field to settle our bodies. So one learns to speak with the tensions that Sally was inviting here. To speak with absolutes or to speak with a sense of binaries is to do some injustice somewhere else.

Our measurements actually perform and reinforce the crisis. So, the question that I'm learning to ask is: How do our solutions reinforce the problems? How does the clinical alliance and the context of therapy reinforce the problems that we're dealing with?

Out of the wound, whether it's climate, chaos, racial injustice, or police brutality, with Deleuzian thought, there are certain dynamics that tend to reinforce a concretion of the familiar. The practices that we swear by hold things in place and refuse us some kind of fugitivity. I speak a lot about fugitivity, about a politics or therapy that is yet to come, that we don't know the contours of, we don't have a language for. '

It's not something new; it's not novel in terms of dismissing what's very vital today, but it's definitely an invitation to be humble. It's a humility; it's an animist humility. It's something that maybe critiques the human. By *human,* I don't mean the individual subject but the territories of acting and behaving, the way we frame spacetime, the way we frame our economies and politics. All of that is called into the room, like Francis beautifully said. And we're being invited to wrestle with that.

So my politics is about how we escape this convergence, this almost toxic cyclicity. How do we break out of the plantation? How do we go elsewhere that doesn't have a name?

I've been doing this work that I call "making sanctuary." It's not a manifesto for a new form of therapy, but it's a speculative fabulation—a way for us to gather around the cracks, the wounds, the stories that want to be told.

The Inuit culture has this ritual called *qarrtsiluni.* The people hunting whales would gather in a dark room and sit with the darkness, waiting for the lyrics of a song to sprout. The word *qarrtsiluni* literally means sitting in the darkness and waiting for something to happen. I feel there is a sense in which we are in a *qarrtsiluni* right now. We're in this place of darkness, but this darkness is productive, it's prolific.

It's how to sit here that is the issue, how we chart a cartography of exile and fugitivity and novelty in the face of resilient imperial structures.

SW: Steffi, you posed us a question before we met today, which was concerning the familiar and how easy it is to slip back into the familiar setting. So I looked "familiar" up. It means "well known from long or close association." You referred to the fact that the familiar is dying, and how do we respond?

The other meaning refers to characters that are rather demonic, which I found very interesting because I would say that there is a tension, that if we're going to survive this moment, we encounter the deeply unfamiliar in our apparent familiarities. If we're not going to do that, I think we're headed toward a very desperate path. So I found the word "familiar" very challenging. Thank you for the way you posed that.

BA: I'd never considered it from that perspective. But yes, my sister, I'm quite wary about what I call psychic gentrification. We colonize the darkness in order to let it be legible to what we already know. We drag the screaming demons from the swamps so that they speak to our needs right now. The eloquence of today has to be a *gasp,* not just another series of words. It has to look something like my son, struggling for language, and me noticing that this struggle for language is not a disadvantage; it's a gift we don't know how to name yet.

FW: It's pretty clear that we've entered into what we could call the *long dark.* It's going to be decades, perhaps generations, in the making before we even get a sense of what might be emergent. But we are in a time of deep descent, and we're leaving the terrain of the self and entering the terrain of the Soul, and we don't have much of a Soul-language; we have a self-language. We don't even have a psychology; we have a self-ology. We've lost psyche, we've lost Soul. So we have to begin to become familiar again with the customs and the manners that closest resemble what Soul is aching for.

SB: One of my questions is whether one-to-one therapy continues to be the necessary focus. Do these times also call for something larger? Do the skills that we bring need a wider context and maybe a different frame?

SW: I think it's not an either/or. There will be a role for people needing to talk to somebody and have therapy. I, for myself, have moved out of that kind of work. This decision was a long time in the making.

We have skills and understanding that can have a different level of application. We can have disavowal in a therapy session. We can have disavowal at COP26. They can have structural similarities, but that doesn't mean the reason for them is the same. How do we recognize the similarities at these different scales?

These are very complicated questions, but I think that's where a lot of our work is needed now, at that sort of level: how to help people face the unbearable feelings that are here. How do we help people, not in one-to-one sessions but in larger groups?

MJR: I definitely think there have to be spaces for all different kinds of things. Personally, I'm too introverted. There was too much difficult stuff for me in the early years to have gone into a group. I tried it, and I couldn't do it. I had to have the one-to-one experience, and I'm sure many other people feel similarly.

I feel passionately that there has to be a space for not just group therapy. There's something else that's called for. It's a kind of coming together to share grief and so many other feelings that people are going through and having difficulty in naming. Unless we can do that, we cannot move forward.

The early years for me, attending Joanna Macy residentials, weren't psychotherapy, but it was very therapeutic. It was life-changing for me to be in the community and witness each other's pain for the world.

Closing with a Dream

MJR: It might seem a bit odd to bring a dream in for my closing remarks, but it's about the death of the familiar.

Just for some context, I've lost both my parents in the last few years, and I've also lost the piece of land where I was born. So it's a lot of loss and a lot of death of the familiar.

In my dream I'm going to visit my father. He was the last one to die. I catch the train and the bus in the familiar way, and I walk from the very small town to where they were living.

But I go a different way than the way I would normally go, and I meet a woman who offers me a cup of tea. She gives me a cabin to sit in, and I'm drinking the tea and I'm planning the weekend. I think about what

I'm going to cook and what I need to buy for my dad, which is what I had done over the last few years.

Then it began to dawn on me that we had sold the house and I couldn't go there. Then I thought, "Where's Dad?" I went through lots of possibilities in my mind as to where he might be. Was he staying with so and so? Was he in a home? Was he living with my sister?

Then it dawned on me that he had died, and it was really shocking, even after all this time. When I realized that the house was gone and that he was dead, my thought was: "Where am I going to go now?" It was really shocking, and I woke up.

It's going to be two years in March since he died. I still feel really shocked. It's as if they have just died.

For me it's not just a story about personal loss; it's how long it takes to take in the death of the familiar, and all the different layers that we go through with the realization of it. It's a story of how long it takes to seep into our bones and how many things we are losing at the moment. Other people have mentioned that it's not just creatures and people— it's loss of trust, loss of familiar ways of doing things, and so on.

I wanted to add how long this takes. We need to slow down. There's a terrible paradox with needing to get on with things so urgently, and yet, we need time to realize and take in the enormity of it. How do we take it in? Losses on such a huge scale. It's hard enough to face personal death. How do we take in the possible loss of the ending of our species? I don't have the answers, but these are the things that are preoccupying me.

Chapter One

From Individual Lens to Cultural Derangement

Transcontextual Reflections on Therapy

Steve Thorp

I am writing here about therapy—or "learning in relationship"—in the context of the trouble that we humans face, and in the face of everything that is being lost.

I will start with Ursula Le Guin's "clear, clean lines" in her essay titled "It Doesn't Have to Be the Way It Is," which speaks to the essential power of imagination: "Why are things as they are? Must they be as they are? What might they be like if they were otherwise? To ask these questions is to admit the contingency of reality, or at least to allow that our perception of reality may be incomplete, our interpretation of it arbitrary or mistaken" (Le Guin, 2017, p. 83).

It is true in the common sense that "it doesn't have to be the way it is." However, our societies, institutions, and cultures seem to operate on the opposite assumption. Humans and other-than-human beings are complex living systems within systems; we are "transcontextual." Yet how we operate and behave seems dependent on "how we do things around here." Attempts to address the ecological and related crises that we face without changing the way things are done will not even touch the sides. And changing the way we do things is infinitely complex, emerging from the very culture that needs to change!

Le Guin (2017) also tells us that the imagination is subversive, and that "subversion doesn't suit people who, feeling their adjustment to life has been successful, want things to go on just as they are, or people who need support from authority assuring them that things are as they have to be," and she is clear that the imaginative act does not say that "anything goes," either within a story or in the context of a life (pp. 81–82).

Yet already there is a problem. The language of therapy carries cultural assumptions that hold us to a "way that things should be." We talk about clients, implying a commercial relationship. We call ourselves clinicians, locating us in the medical realm rather than the imaginative. We talk about healing, mental health and illness, consulting rooms, transference, and treatment plans.

Psychologists and psychotherapists regard ourselves as open-minded, contextually informed creatures, but we too are subject to the boundaries and constraints inherent in our cultures, and we reflect these (often unthinkingly) in our practice. The world we are troubled in is one in which the dominant culture seeks and promotes straight-line solutions to isolated problems and pathologies. Therapy tells us there is something wrong inside (and it will be *inside*) that can be adjusted, worked through, and healed. Yet this sounds like the kind of "mental mono-cropping" that Nora Bateson (2016) cites as indicative of a language and culture that "favor singular focus, clear definitions, and linear narratives of causation" (p. 19).

In dominant Western culture, therapists and their "clients" live with crisis and collective trauma, cultural assumptions, and straight-line solutions. These cultural phenomena inevitably emerge in the work we do. We do not choose to reduce the complexity of our world to straight lines and determinisms, but somehow we do. We do not regard "healing" as exclusively individual—recognizing that suffering emerges from dysfunctional systems—but we find our way back to one-to-one work with the goal for the individual to get "better," following an upwardly developmental model of progress that is in itself deeply flawed.

Therapists who want to imagine another way are caught in a dilemma: wanting to go beyond the internalizing, decontextualized, reductionist, and deterministic aspects of our professions, but getting our validation, accreditation, "professional development," and sense of community from our work *within* these boundaries and codes. And clients often arrive to see us with the same cultural expectations we have about how we might heal them or take away their suffering.

James Hillman's work (see, for example, Hillman, 1996; Hillman & Ventura, 1993) sets the sadness and madness of individuals in the wider social, ecological, political, and mythical realms of the world, stressing that the relationship between inner and outer is ever-shifting, complex, and imaginal, but also that therapy has historically failed to reflect this rich interdimensionality.

So how can we as therapists work with troubled individuals in a world that is experiencing trauma, disaster, disconnection, and the possibility of ecosystemic collapse, and *not* work with the dimension of the ecological self? How can the pain of the world built through colonialism and late-stage carbon capitalism *not* be internalized and emerge as pathology, eco-anxiety, or other symptoms of trauma?

For clients "racialized" and "gendered" within these systems of oppression and trauma, the need for such therapeutic "contextualization" is essential, with all that this implies for our practice. Feminist and anti-racist writer Carol Gilligan (2002) reminds us that the subjective psychological experience is revealed in culture that "appears in the unspoken; it is the way of seeing and speaking that is so much part of everyone's living that it never has to be articulated. Culture is revealed in those moments when someone does not know how they are supposed to see or to speak, or when the tacit rules are broken" (p. 86).

If deep conversation and relational learning can be therapeutic, and if imagination points to new ways of being, where are the spaces in which we can explore our place in the world without being tied into these unspoken cultural traps and dilemmas that always threaten to catch us? And, furthermore, might it be possible for relational, contextual, therapeutic learning to take place in a one-to-one conversation—in individual therapy—when this is a space that is often replete with hubris, appropriation, determinism, hierarchy, and exclusion?

Group and family therapy suggests that the content of a conversation is not necessarily the key to its vitality. Something happens "between" the people who make up the system. Their contexts might seem invisible but are implicit in their relationships and can coemerge in ways that result in mutual learning and vitality.

Since the beginnings of family therapy in the 1950s, practitioners have been concerned with reembedding "troubled" individuals into wider systems and exploring how people respond in relation to their multinodal contexts. Some of the groundwork for this was undertaken by Gregory Bateson and his colleagues, who were bringing ideas like "ecology of mind" and the "double bind" into the wider culture.

His daughter, Nora Bateson, has built on these themes. Like Le Guin, Bateson is not looking for easy answers. She tells us that while big, "cold" data can tell us something about the state of things, other kinds of knowledge can only exist in constantly shifting relationships, in ecologies that respond and adapt in complex interdependencies. She describes this kind of information, which she calls "Warm Data," as being "wiggly, unpredictable, and sometimes invisible. In this way . . . the family, the biosphere, and society, continually calibrate their relationships. . . . In the Warm Data work, I have seen the most beautiful and unexpected shifts. These shifts are so full of possibility that they gush in like necessary oxygen after holding one's breath" (Bateson, 2022a).

More recently, Bateson has turned attention more directly to "healing." In her essay "Tearing and Mending: Transcontextual Learning and Healing" (Bateson, 2022b), she writes: "Healing is dispersed, it is not only in the spot that hurts. It takes many contexts shifting to support a change in the identified problem-zone. A good friend, long walks, fresh food, music, soft blankets . . . and ideas that change. Perception changes in small flashes, watery seepings, frantic scribbles, savored flavors, long silences and funky beats" (para. 11). These ideas and practices are counterintuitive in the clinical worlds of boundaries and outcomes; however, many therapists *do* understand and recognize them in our work. Often, we do *not* aim for outcomes but simply allow "Warm Data" to flow in, through, and between the stuff of the conversations we happen to be having with our clients. And sometimes—somehow—something like "healing" might occur.

I have become intrigued by how transcontextual spaces might hold the relational conditions for the necessary "adaptation" to times in which "the familiar is dying." There is more to explore, but I am clear that the work of therapy (in this paradigm) is an act of imagination, relating, and mutual learning. Moreover, it takes place in spaces where not just anything goes. Like writing from the imagination, the work requires rigor, presence, and a clear eye for what is.

For as Ursula Le Guin (2017) reminds us: "Down on the bedrock, things are as they have to be. It's only everywhere above the bedrock that nothing has to be the way it is" (p. 84).

Psychotherapy at a Cultural Threshold

Chris Robertson

A common response from participants in workshops who have awakened to the climate crisis is: "What should I do?" While there are many activist organizations, my emphasis here is on staying with the troubling feelings that come with the awakening. These are sick-making feelings. It is no wonder that participants might want to disperse such feelings through engagement with actions that counterbalance complicity in the industrial despoiling of the planet.

Staying with feelings is a familiar psychotherapeutic strategy but here with a different twist: these feelings are *not* just personal, left over from family pathology. They are constituents of our cultural malaise—a hyper-individualism that dysfunctionally focuses on individual achievement and responsibility. I will be exploring how to listen through individual symptoms to the unspoken, unseen, and uncontained pathos of our culture. Listening through implies a porosity—a permeability of both individual defenses and the walls of our consulting rooms.

This paradigm shift in psychological focus from the individual to the culture, from personal growth and individuating to decentering and kinship, comes with many challenges and dangers. The shift goes against the grain of normative psychotherapy's attempt to fix things and opens a radical unknowing. Such a shift amounts to a collective rite of passage, catalyzed by the incipient environmental crisis.

Moderns, those of us living with the privileged entitlement of industrial exploitation, have defensively failed to change our collective relation to the other-than-human and our planet. Modernity mandates progress and power-over. Modernity requires psychotherapy to fix or cure what is "wrong," often making it complicit.

In this writing, I speak for making space for the wounded, broken, marginal, inferior aspects of the collective psyche through which a profound healing of our estranged state could emerge. I attend especially to the humbling difficulties of being with unbearable feelings of loss, tragedy, and despair.

Rites of Passage

In 2008, I ran a leadership workshop on a remote Swedish island where we explored dying as a rite of passage that brings with it deep anxiety and potential transformation. As I wondered how to start the training, I suddenly saw an eerie parallel between the Swedish island, with its nineteenth-century lighthouse, and Alderney, also a small island with an important lighthouse, where the previous week I had gone out on a fishing boat to scatter my late mother's ashes. My skin tingled with the awareness of this coincidence.

It prompted me to tell the group about the death of my mother. Initially, I had wanted to keep this private, so I was surprised to receive deep resonance from the participants to the connection between death and leadership.

Such resonances occur when vulnerability, in this case mourning the loss of my mother, dissolves the boundaries between the individual and others. The death of my mother was a personal threshold with personal grief, and it was tinged with a collective sense of tragic loss. The collective loss needs a hook into the personal to be experienced and inflates it with nonpersonal energy. This can be confusing, and it is important to distinguish the collective overwhelm from the personal. When the two come together as in the coupling of climate catastrophe and social upheaval, this can feel overwhelming.

Rites of passage, whether explicit such as in weddings and funerals, or implicit such as in midlife, function as potential renewals in which the unknown penetrates us and our life may never be the same. It is as if we moderns are facing just such a rite of passage, not as individuals but as a culture, in which some basic assumptions about our security and trust in the future are challenged. This is the unraveling of modernist culture, a dark time when things fall apart and there is little or nothing to hold onto.

Just as my personal ego was broken open by death and tragedy, so too is a prevailing cultural identity with facing potential ecocide. A cultural identity is an abstraction in our own image through which every experience is filtered via social scripts, mores, and implicit belief systems. It imposes its control through power structures and dominant narratives that seek to protect and maintain business as usual. This is where creative innovations flounder within a systemic miasma of threatened social defenses. Our very language has been wounded by our culture's demands for efficiency and precision that forsake rich ambiguity.

Extracting ourselves from this co-constructed reality, which has shaped not just our social world but also our beliefs and perceptions,

may feel like a betrayal of bonds of loyalty that tie us to our social reality. Attempts to leave the privileges and exceptionalism of Modernity can
activate social defenses and cultural taboos. It is as if the modern human
unconscious has been colonized, and attempts at decolonizing create
waves of collective fear, distrust, and reactive panic.

Facing into the reality of this fear and not wishing it away with escapist
hopes and promises is a huge cultural challenge. In a 2014 workshop on
"Radical Hope," I invited participants to remember a time in which they
had felt unable to think of any solution—a time with no future or an empty
future.

We dropped into the presence of the unspeakable: the dread of the
void, the abyss. Strangely, the sharing of these horrors touched into poi-
gnant intimacy. We witnessed a falling apart together. A spontaneous
group sculpt emerged that allowed an embodied deepening; a melding
of the grief and pain at our losses with a transformational sense of being
held; a new possibility of meaning.

It is through facing the fear that transformation becomes possible.
Perhaps psychotherapists need to re-vision thinking about catastrophe
not as aberrant or pathological but as valuable learning. Susan Kassouf
(2022) suggests an imaginative extension of the infant-mother trauma of
"no-breast" to the sixth great planetary extinction, "not no mother, but
possibly no human species" (p. 3). She writes:

> A traumatized sensibility has learned from experience that annihila-
> tion is thinkable. It can bear the tragic, the feeling of irreparable bro-
> kenness, or in environmental terms, that we have entered a time of
> post-sustainability. There is a living memory, or awareness, or ability
> to imagine the inevitability of repeated collapse, social and otherwise,
> violence and war. And yet, an ability to think and translate think-
> ing into action in a state of precarity is maintained. The world is not
> assumed to be safe.
>
> Just as annihilation is thinkable, despair is bearable. A traumatized
> sensibility does not resist but can tolerate despair. (p. 11)

Instead of a stoic resilience in the face of loss and eco-anxiety, we could
imagine psychotherapists enhancing their traumatized sensibility. Instead
of a fragile holding in the face of the fallout from the cultural unraveling,
our own wounding could enliven our capacity to be with the suffering.
This is a transformative shift that comes with facing into the trouble and

deeply accepting the discomfort that comes with that. It is a shift from corrective balance to humility and from bitterness to the salt of wisdom.

One recent morning I faced a challenge when, upon getting up, the room started spinning and I had to drop onto the floor just to avoid fainting. I was floored. Brought down, frightened. It turned out my heart was on a go-slow. I had not consulted it when offering workshops on heartbreak. In one such workshop, I said the following:

> We will be exploring how to bear these challenging feelings often disowned, denied, and split off in our cultures. We may touch on the dangers of vicarious trauma in bearing witness and the subtle disconnections that unconsciously communicate a no-go area.
>
> These feelings can threaten to overwhelm, emphasize a helpless loneliness, or trigger traumatic memories. Their challenge can also catalyze a shift away from personal suffering to a planetary suffering, a pain of a world severely damaged by our way of life. This was what Glenn Albrecht recognized in wanting to name that kind of homesickness caused by environmental violation—*solastalgia*.

During the workshop, I asked: "What is it that we find unbearable in facing the climate crisis?" The experience of falling to the floor, as a possible failure to process the collective heartbreak, exposed my vulnerability directly. But this "being-with" the experiences of heartbreak around the climate crisis is not simply personal. It is also a rupture in the fabric of our lives, our cultural norms. The prevention of such ruptures maintains the social safety net that has insulated many of the Western democracies. When ruptures such as food scarcity or wildfires happen, they create a semiconscious panic, which activates collective defenses. Cultural identity can feel hopeless in the face of deeper forces outside its control.

When normal procedures have failed, something different is required. Culturally we are at an impasse where cognitive reason fails. Clearly, some collective unconscious processes are at work. Can psychotherapists help with the passage through this cultural threshold and the imperfect practices of bearing what feels unbearable?

Imperfect Practice: Healing through Humility

The simple answer to the question of whether psychotherapists can help with collective distress is "with difficulty." In supervision, many psychotherapists report feeling helpless in relation to climate distress, as if that were an incapacity. But while the helpless feelings may be failures of "normal" practice, they represent spaces where the "normal" has broken open. Bayo Akomolafe, one of the contributors to the group conversation in this book, speaks of "generative incapacity" to indicate the apparent paradox of the helpless stuckness that brings vitalizing healing. This is not the helplessness of a victim constellated around depressive states, or a learned helplessness developed to defensive advantage. These are practices of humility learned through failure and being with the "dying of the familiar."

While writing this chapter, I received a warning dream. In the dream, I was living in a hut at a campsite. I heard fractious voices outside complaining about someone writing about other-than-humans. Clearly, they were talking about me. I went outside to find my neighbor already engaged with two strange-looking men. I said it was me that they were looking for and nervously asked what they wanted. I could not really hear their answer and looked to the neighbor, who said, "I cannot translate for you, as I am human too."

On awaking from the dream, it dawned on me that the visitors are not human and that I am caught within a human-centered perspective. Perhaps I am overreaching in my intention to bridge an interspecies divide? Staying with the impasse presented in the dream is awkward. I want to find a way through, some crack in the edifice of our human exceptionalism that would allow me to escape this alienation.

This uncomfortable tension is familiar to those attending to thresholds. How to be present to the uncertain liminality, attending to the discomfort while inhibiting our mind's attempts to escape?

In many supervision sessions, psychotherapists acknowledge this impasse. They feel as if they ought to get somewhere with their client, and when this doesn't happen, they read it as a failure. They report feeling a helpless incapacity, a feeling of being stripped of their sense of agency.

Thresholds can evoke vulnerability. We are transiting between worlds, between social structures and cultural frames. The process is disorienting, and not to feel vulnerable in this process implies a dissociative defense.

The collective shock at climate disasters and social breakdowns is not outside the consulting room; it permeates the consulting room. To acknowledge and give permission for these nonpersonal troubles to be voiced, psychotherapists need the humility to relinquish normative practice.

One of the gifts that psychotherapy has to offer is a quality of attention that is in short supply elsewhere in our culture. Instead of giving this quality of attention solely to individuals, what if it were developed as a cultural practice within the wider community of beings? In this shift of psychological attention from individual focus to that of the culture, I distinguish three layers:

+ The need to make the walls of the consulting room like semipermeable membranes that offer a porous containment to the troubles of the world.

+ The need to read the signs of cultural unraveling as part of a necessary and vital change to Modernity.

+ The need to bring transformative practices into community.

Listening through to what is beyond the walls may be less obvious, as there are years of psychotherapeutic history privileging the inner container. In Hillman and Ventura's (1993) provocatively titled book *We've Had a Hundred Years of Psychotherapy—and the World's Getting Worse*, they challenge implicit social assumptions about improvement and progress. If society itself is sick and the consequences of this appear in consulting rooms, perhaps an endeavor to treat individuals is collusive? Are we psychotherapists caught up in a collective trance of improvement—a heroic endeavor to triumph over our culture's problems?

The story of triumph is reflected in describing our current geological epoch as the Anthropocene, which can be read as ignoring how our human culture is sustained by the very ecosystems that moderns exploit. In contrast, becoming aware of our human responsibility for the erasure of other species and how this threatens the sensitive ecological balance on the planet is salutary and sorrowful. In an interview (Milstein & Castro-Sotomayor, 2020), David Abram said:

As soon as we notice the humble, earthly ancestry of the word "human" then another possible name for this new geological epoch immediately suggests itself. If we really wish to underscore the human species as a key—if unwitting—perpetrator of this new and rather calamitous state of affairs, wherein so many other animal and plant species are tumbling into the oblivion of extinction, then why not call this epoch the *Humilocene*. The Humilocene: the epoch of humility. That does emphasize our species' outsize influence right there in the name—Humilocene—yet it also feels awkward and disturbing, at first, for it carries an echo of another word that shares the same origin, which is "humiliation." (para. 21)

In my psychotherapeutic work with men, I have noticed that their experiences of encounters that were humbling are often described as humiliations. This might have been a challenge to their competence at work or a relational skirmish in which they felt exposed. While acknowledging that the experience was hurtful, we often explored the potential value of the encounter if they could get past their feeling of humiliation. This distinction between humbling and humiliation is significant at the cultural level.

The Humilocene, a potential epoch of humility including remorse in coming to terms with past abuse, is an apt context to understand the challenges involved in the rite of passage for the end of Modernity. This cultural context provides a wider and deeper frame within which psychotherapists can listen through beyond individual stories. This is not so much a competence in cross-cultural perspectives as a shift away from an overemphasis on the individual.

Psychotherapy offers ways through the cracks in our constructed reality. Winnicott (1971) attends to the vital importance of play and emphasizes the function of the imagination in the construction of reality. In that delicate but richly creative liminal zone between inner psychic reality and the external world, Winnicott recognized the function of *transitional objects*. These may take the form of a security blanket or cuddly toy that stand in for the absent mother. But what of cultural breakdowns?

All ruptures need holding spaces to be encountered. But where do we find these if the culture's roots are rotten?

The costs of climate change are felt in our bodies as much as in our despoiled land, acidified oceans, or poisonous atmospheres. They are experienced in our psyches in dreams as much as a literal environmental degradation. Our psyche-soma sensibility is embodied. Our bodies are porous and fragile, subject to infection. Our skin is a permeable boundary that breathes. Wounds and scar tissue remind us of our sensitive connection. This embedded entanglement is what allows our senses to apprehend the world. Much as the ego might want us to believe in our separate independence, we are interconnected with all living cells.

There is a reciprocal affinity, as Abram (1997) puts it, between our flesh and the flesh of the planet's biosphere that comes preattuned through having coevolved together.

Complex Transformation

Psychotherapists can serve as catalysts in a holding transitional space as old forms break open and in supporting the digestion of collective grief and remorse. These spaces allow the experience of overwhelm, helplessness, and despair to be witnessed without the need to fix them. This is how we can digest the horrors of what our modernist exceptionalism has done without activating a defensive shutdown.

To the extent that psychotherapy brings transformation, it is largely because it allows what was unspeakable to be spoken and what was previously unbearable to be experienced and thought about. If these collective emotions are understood as personal, they feel beyond us as individuals, and they are. Individual complexes such as a persecution complex, a martyr complex, or a guilt complex can help us understand our repeated, often delusional patterns of behavior as stemming from a condensed emotional root (or complex). If we can make the shift away from the traditional emphasis on the individual and toward what is shared in the culture, from personal fears to collective ones, from individual complexes to cultural complexes, we may feel not only less burdened but also less isolated.

With the notion of cultural complexes, I refer to Jung's notion of *complex* as developed by Tom Singer and Sam Kimbles (2004). They speak of group complexes, which collect experience, especially of trauma. Complexes function through inhibiting, forbidding, and denying emergent realities that are unconsciously experienced as threatening. These historical group experiences have taken root in the collective psyche, and when they are activated, the group identity is activated to defend against a part projected out onto a suitable hook. The sole function of a traumatized cultural defense is safety and survival.

Sam Kimbles (2014, p. 11) writes that cultural complexes shape the collective convictions of group life: "These patterns, at the group level, fall into place around the instinctual needs to belong and have an identity via identification that pushes for recognition. . . . Potentially gaining awareness of cultural complexes allows us to recognize our own subjective responses to the broader social situations of which we are both part and participant."

This potential space is an open space for transformation. Having come through what was felt to be a catastrophe that broke us open, we are knowingly permeable and familiar with that frightening relinquishing of control. This letting go potentially catalyzes a regeneration of our being, a being-with and belonging-to rather than alienated isolation. This symbolic rite of passage can vitalize our capacity to be in an unsafe and uncertain world without closing off defensively into a privileged huddle.

Through my own midlife rite of passage, I began to apprehend clients' presenting issues not as developmental failures but as a stumbling upon an unknown threshold. It was as if they kept going through the same familiar door, unaware of the portal to an as-yet unlived life. Such blindness has many parallels with the denial of climate change. Obsolete certainties may cling to us like blinders where the uncertainty of the unlived and unknown holds terror.

Ending Reflections

I started by speaking of sick-making feelings. In my own case, these are lovesick feelings of desolation that have tenderized me. It makes me weep facing the calamity of collective losses—places, people, and species that have been swept away in the mad rush of progress.

And this tragic damage may mark a threshold for a rite of passage. We can no longer rely on the Earth to look after us. We either perish or go through a symbolic death as a rite of passage. I have explored how psychotherapy, as a modern human endeavor, is itself at just such a threshold in a potential move away from fixing problems to being with or staying with the troubles of our time.

Managing better may not be effective in bringing cultural change. Modernity cannot tolerate "failure." Respecting and giving a place to failure as incapacity is suitably humbling. Sometimes not-doing is more powerful than any action. Inhibiting familiar practices, stopping an overly busy life, and relinquishing privilege are also agents of change.

Paradoxically, deep acceptance, not to be confused with passive resignation, opens a portal to potential reparative healing and to the Humilocene—a potential epoch of humility. This may be the dangerous threshold across which our modernist culture transits, a threshold that demands a frame within which disappointment, failure, sadness, and tragedy are given a place.

I suggest that the vulnerability of the culture can be the portal through which transformation occurs. In exploring the vulnerable, often seen as weakness, I hope to have given more of a place to aspects of the cultural shadow—those inferior qualities that often reconnect humans to their animal and animating nature.

This approach accepts the inevitability of the current cultural unraveling, not with reluctant stoicism but with grief, humility, and an open sense of wonder about what is emerging through this cultural rite of passage that marks the end of the world as humans have known it. Healing here is not a normative return, a palliative health-and-safety contrivance, but a risk-laden passage through unmarked doors. We are being invited to practice going against a normative grain that maintains things as they are. The threshold invites a fertile abandonment of control to be welcomed by the stranger who is ourselves.

Chapter Two

Inviting the State of the World into the Consulting Room

Why Aren't We Talking about Climate Change?

Defenses in the Therapy Room

Trudi Macagnino

There is increased awareness and concern regarding the climate and ecological emergency (CEE), with 75 percent of people in Great Britain stating they are worried about climate change (ONS, 2021). Attention is now being given to the psychological and emotional impact of the CEE. The latest Intergovernmental Panel on Climate Change report mentioned the impacts on mental health for the first time: "Mental-health challenges, including anxiety and stress, are expected to increase under further global warming in all assessed regions, particularly for children, adolescents, elderly, and those with underlying health conditions" (IPCC, 2022, p. 15). Evidence supports this statement, with depression, PTSD, guilt, grief, anger, and anxiety being associated with climate change (Lertzman, 2015) and being particularly prevalent in young people (Hickman et al., 2021). These symptoms are often referred to under the overly simplistic umbrella terms of eco-anxiety or climate anxiety. These labels have a clinical ring about them, although many believe they are a healthy response to the CEE, rather than a pathology (Bednarek, 2019b).

Psychotherapy and counseling professions have begun to recognize the need to attend to eco-anxiety. Professional organizations such as the British Association for Counselling and Psychotherapy, the UK Council for Psychotherapy, and the British Psychological Society have run continuous professional development (CPD) events and issued special editions of their journals (UKCP, 2016; Wainwright & Mitchell, 2020; Brown, 2021), and some have even made statements declaring a climate emergency (RCP, 2021). However, as an integrative psychotherapist working in private practice, I have not seen clients coming to sessions wanting to talk about the CEE, despite the evidence suggesting that eco-anxiety is increasing. This was a starting point for my PhD research.

Through interviewing therapists and clients for my research, a puzzle presented itself. In therapy sessions, clients and therapists easily and comfortably talk about the natural other-than-human world, such as animals, pets, the natural environment, and landscapes. They speak about the healing and restorative powers of Nature, of special relationships with other-than-human entities, and the way in which the natural world is part of their spiritual practice. Despite the obvious value placed on the natural world by both therapists and clients during their work together, the CEE is rarely spoken about directly. Therapists told me that sometimes, however, clients make side-mentions related to the CEE, such as noticing unusual weather or referring to pro-environmental behavior, such as going plastic-free. In these cases, therapists typically choose not to explore the topic further, instead seeing these comments as small talk. In parallel, clients also choose not to say more or go any deeper into these areas. So the puzzle for me was: if the natural world is so important to clients and therapists, why is the CEE—a threat to something they love—not being brought into the room?

The interview data suggested that this was not due to a lack of care or concern about the CEE. It is more likely that eco-anxiety caused by confronting the reality of the CEE is potentially an overwhelming experience of existential magnitude (Lertzman, 2015). It is, therefore, not surprising that we are all likely to defend against it to various degrees at times. Sally Weintrobe has written extensively about disavowal, the unconscious defense of turning away from the reality of the CEE (Weintrobe, 2013). She has also suggested that much of this disavowal is social, a result of our Western individualistic culture (Weintrobe, 2021). Such a psychosocial perspective sees that the individual is to be found in the social and the social within the individual in a mutually constructing dynamic, not all of which is conscious (Frosh et al., 2003).

In this chapter, I outline a psychosocial explanation for the puzzle and an understanding of the unconscious defenses at work in the therapy room that serve to keep the CEE out. I specifically draw on concepts of containment (Bion, 1962), transitional space (Winnicott, 1971), thirdness (Benjamin, 2004, 2009), and social defenses against anxiety (Menzies Lyth, 1960) to make sense of my findings.

Method

Taking a psychosocial position concerned with both inner and outer worlds, the psyche and the sociocultural, I utilized psychoanalytically informed methodology that recognizes hidden unconscious structures and defenses that lie "beneath the surface" and "beyond the purely discursive" (Clarke & Hogget, 2009, p. 2). Specifically, free association narrative interviews (Hollway & Jefferson, 2000, 2013) were used with therapists, and the biographical narrative interview method (Wengraf, 2001) was used with clients. Each participant, apart from one ("Lester"; see table below), was interviewed twice. Due to the COVID-19 pandemic, 24 of the 27 interviews were conducted online using Zoom live video. I kept a research journal throughout the project to capture my thoughts, ideas, feelings, dreams, and reflections, and the journal's contents became important reflexive data.

Seven therapists were recruited through convenience and purposive sampling (Patton, 2002). The only inclusion criterion was that participants had to have an active therapy practice.

THERAPIST PSEUDONYM*	GENDER	APPROXIMATE AGE	MODALITY
Jenny	Female	50s	Core process
Lester	Male	60s	Humanistic/ integrative
Lee	Female	60s	Gestalt
Nellie	Female	60s	Integrative
Amanda	Female	50s	Core process
Sarah	Female	50s	Person-centered
Rob	Male	30s	Dramatherapy

*All names have been changed to ensure anonymity.

Seven clients were recruited through a similar sampling method. Clients were unconnected with the therapists. Inclusion criteria were that participants needed to be in or have recently completed therapy and were sufficiently emotionally resilient to take part, as judged by their therapist.

CLIENT PSEUDONYM*	GENDER	AGE
Sean	Male	45
Margaret	Female	75
Deborah	Female	30s
Phil	Male	39
Elaine	Female	64
Martin	Male	62
Helen	Female	43

*All names have been changed to ensure anonymity.

Pen portraits were written about each participant, giving background and context, and capturing individual characteristics, the emotional tone of the narrative, the core story line (Cartwright, 2004), and my general observations and impressions. This process served to capture counter-transferential data that was then incorporated into my analysis.

Thematic analysis (TA) (Braun & Clarke, 2006) was then conducted for each individual data set (therapists and clients), generating codes, themes, and patterns. Specifically, reflexive TA (Braun & Clarke, 2019, 2021) was used, which puts "the researcher's role in knowledge production . . . at the heart of [the] approach" (Braun & Clarke, 2019, p. 594). This is consistent with a psychosocial methodology, where the researcher is understood to be co-constructing the narrative. A more focused reflexive and psychosocial lens was used by further interrogating the data with questions designed as prompts to deepen my analysis still further.

Psychosocial research assumes both subject and researcher to be anxious and defended (Hollway & Jefferson, 2013). Therefore, even with reflexivity it is not always possible for the researcher to access their own unconscious processes. I, therefore, involved colleagues, supervisors, and peers in the form of research panels to gain differing perspectives on extracts of the data that I found particularly interesting, confusing, or strongly indicative of a particular theme.

The final stage of analysis was to consider the two data sets of therapists and clients together as a whole, identifying unifying themes as well as any differences.

Findings

As already outlined, one of the key findings was the contradiction between the importance placed on the natural world by therapists and their clients, and the relative lack of direct reference to the CEE by them in the therapy sessions; this was the puzzle. The explanations for this puzzle constituted the remainder of my findings and were captured under two broad themes: "what therapy is for" and "feelings and defenses."

THEME: "WHAT THERAPY IS FOR"

Therapists drew on a discourse of professional and ethical practice emphasizing the importance of not bringing their own agenda into the therapy or of leading the clients in a particular direction. They used the therapeutic contract as an explanation for not exploring clients' feelings regarding the CEE in a deeper way. Jenny, for instance, does not open up side-mentions about the CEE from clients because she is "more listening out for what that person is *really* wanting to bring in that session" (emphasis added). I sensed that Jenny was justifying this when she stressed that this would be "in keeping with what we've contracted to work on." The need to justify suggests some anxiety around this. The disregarding of the side-mention seems to me a defense, an avoidance of stepping into material that she would rather leave be.

In a similar vein, Lee said, "I don't want to be directive or leading or, you know, I'm not that kind of a therapist." The suggestion was that the "kind of therapist" who leads clients, in this particular aspect at least, is an unethical one.

In this way, therapist and client explicitly and implicitly agree on the therapeutic contract—in other words, this is what your therapy is for—and in so doing they keep sociopolitical concerns such as the CEE out, as though external contexts have no impact on the internal world.

This focus on the therapeutic contract is then continued and further reinforced in supervision, as this quotation from Jenny demonstrates: "With my own personal supervisor . . . I would tend to be focusing on the really, sort of, psychotherapy part of the relationship with my clients with him . . . attachment patterns, looking at people's own personal history, looking at people's thought processes. Um, looking at my relationship with them, you know, just me and them in the room, you know, without that bigger picture, so it's like . . . the micro levels of the relationship."

Similarly, client participants saw the purpose of therapy as dealing with their personal issues, the small stuff of everyday life, such that large-scale global issues such as the CEE are not relevant to the task of therapy. Martin described the topic of Nature and the CEE as "highbrow" and said one needs to "get the foundations" sorted out first, meaning personal issues. Elaine said she does not want to talk about it with her therapist until she's "got the foundation." I think what Martin and Elaine meant by "foundation" was a sense of personal resilience, a firm base, which they need before the feelings regarding the CEE can be explored, perhaps because they sense that CEE-related feelings are potentially of a different order of magnitude, whereas the small stuff is stuff they can do something about. When talking about the "small stuff" they are alluding to their own sense of smallness and powerlessness in relation to the enormity of the CEE.

The shared view about therapy from both sets of participants is that therapy is for the microlevel issues of self and relationships with other humans. These issues are agreed upon between the client and the therapist and then between the therapist and their supervisor. In this way, a parallel set of processes is established, both explicitly and implicitly, in a strongly held therapeutic frame.

The adherence to a therapeutic model or framework seems to provide some sense of structure and security for therapists, a place to retreat to perhaps. For example, Nellie explained how she will "fall back" on her core model; she knows she can "go there" and "do that" when she does not know what to do in a session. This falling back suggests to me a kind of collapsing into, a retreat from something threatening, rather than a stepping into the space of meeting the client, of not knowing and allowing oneself to be impacted by the same things as the client.

The setting of the therapy also appears to play a role in the avoidance of exploring clients' potential anxieties regarding the CEE more deeply. For therapists working in organizations, the goals of the organization also serve to direct the therapy. This was expressed particularly strongly by Nellie, who works in a National Health Service Increasing Access to Psychological Therapies service. The focus on outcomes, measuring symptoms, and evidence-based practice serves to restrict the therapy, making Nellie's work "narrow," in her words. She spoke about the pressures to keep sessions to a minimum: "So I think there's been more pressure to keep the sessions to the sort of minimum that we need in order to get clients to a certain stage of recovery [laughs]. So I guess that's, you know, been a performance pressure, like, oh, well, I've talked about *x, y,* and *z,* and they seem to be better now so that's worked, job done, you know, out the door [laughs]."

Although Nellie felt frustrated, organizational goals allow her to shift the responsibility of what to focus on in the sessions from herself to the organization.

There was a general assumption among the clients that people, their therapists included, do not want to listen to them talk about the CEE. Clients spoke about being met with a roll of the eyes by people and that conversations were closed down very quickly. This social construction of silence around the CEE (Zerubavel, 2006) leads to a general hesitancy to explore the subject and a vigilance about others' potential reactions. This even played out in my interviews. Clients made apologetic comments as they explored their feelings about the CEE:

"I'm sorry, this is a bit deep" (Martin).

"[Sorry, I'm] being too serious" (Margaret).

"Oh, I wonder how you're feeling? Maybe you're gonna go away feeling really depressed. And thinking, Deborah's a real Debbie Downer today [laughs]" (Deborah).

I think participants were concerned that I would see them as a "party pooper," someone who brings people down. I suspect this is how others have made them feel, and this has contributed to a reluctance to even talk to their therapist about it. Martin confirmed this when he said, "I'm thoroughly enjoying talking to you about it, because you seem to understand where I'm coming from. But if I was to have this conversation with . . . another counselor, they wouldn't necessarily be, I wouldn't have a connection, they won't understand." Martin felt met and heard during the interview and less alone with his eco-anxiety.

Helen described her therapist as "cerebral." When she tried to talk about her grief regarding the CEE, her therapist interpreted it as a projection of her grief for her mother, who had been absent for much of her childhood. This was the therapist's focus on Helen's internal psychological process rather than on the external reality of the CEE, which is common (Jackson, 2021) and is linked to the idea of what therapy is for. However, it can also be seen as an unconscious avoidance by the therapist of opening up material which they themselves may also find distressing.

Client participants seemed to take responsibility for what was talked about and not talked about in their therapy sessions. This tendency is clearly shown in this extract from Sean, as he contemplates why he has not spoken about the CEE with his therapist: "I know it's my responsibility, because counseling sessions are—you know, you've got a helper and a helpee, and, er, you know, the helpee brings everything to the whole context of the conversation. . . . So I know it's my responsibility that I haven't brought enough about how I feel about that, how it affects me."

My experience of how therapy sessions unfold, however, is that therapist and client are both jointly involved in the process of deciding what to talk about. This taking responsibility had an echo of protectiveness to it. For example, Phil wanted me to know that his therapist was really clever at "unpicking" things and "joining things together." Deborah emailed me after the second interview and asked me not to name her therapist in any published material. However, who was being protected from what would need further exploration? For instance, was the client protecting their own idealized version of their therapist? Did they fear that the CEE was too dangerous a topic to introduce, something that could disrupt the therapeutic alliance? Did they fear a withdrawal of their therapist's approval if they introduced the subject? Were they protecting the therapist from feeling distressed themselves about the CEE? Such possibilities could be worked through in the transference and links made to any reenactments of past experiences, such as having to protect a parent.

THEME: FEELINGS AND DEFENSES

Data from the interviews suggest to me that feelings regarding the CEE were present for clients and therapists, but these feelings were defended against in unconscious ways. These feelings were often beneath the surface and arose in participant dreams, through interpretations I offered to participants during the interviews and in my own countertransferential responses during the interviews themselves and subsequently when analyzing the transcripts. I will focus here on the predominant feelings and defenses apparent in the narratives.

Guilt was often expressed as an individual sense of not doing enough but also as collective guilt, e.g., "We need to change our ways" (Sean). Participants often emphasized their attempts at being green, such as recycling and avoiding plastic. I see these efforts as attempts at making amends, of easing the guilt.

Grief was expressed, and it was associated with loss. For example, Jenny told me about an incident when she accidentally vacuumed up a tiny feather gifted to her by a client: "I had this real turmoil between, do I risk my awful dust mite allergy, which will send it off for the next week, by actually trying to look through to find it? Or to just accept that maybe it's gone. It felt big, it felt like a big thing. I felt really sad that that had happened."

Jenny's sadness could signal some unconscious grief about the destruction of our fragile natural world. Her turmoil alludes to the dilemma we all face in relation to the CEE: do we act and inconvenience ourselves, or accept the loss of species, land, and people? As Jenny said, it is a really big thing.

Fear was apparent in the narratives through allusions to war, famine, and death, and fear also surfaced in participant dreams. The therapists tended to speak about their clients' anxieties rather than their own—a way of unconsciously projecting their own fears onto others. Advising clients not to listen to the news was a common way of avoiding facing the reality of the CEE, as in this quote from a therapist: "Even if that really is going on . . . if I'm not reading the news, if I'm not listening to all this negative talk, my life actually hasn't changed" (Sarah).

One of the most common unconscious defenses evident in the narratives was that of splitting. The CEE was treated as separate and distinct from Nature as a nurturing and healing resource, and participants did not move spontaneously in their narrative between these topics without being prompted by me. Participants spoke about their relationship with the other-than-human as being easy and unproblematic, whereas relationships with people are difficult. What is split off here and kept out of awareness is the fact that our relationship with the other-than-human *is* problematic. It is extractive, exploitative, lacking mutuality, and destructive. In this way, participants split off their "Nature-destroying self" from their "Nature-loving self."

Similarly, references to Nature as dangerous and threatening were notably absent from the narratives. No one mentioned recent natural disasters such as mega-bushfires, floods, hurricanes, and droughts, all of which have caused devastation and loss of human and nonhuman life. Nature as dangerous and threatening is thereby split off from a benevolent Nature. Participants were able to retain their idealized view of Nature while the natural world as destructive and savage stayed out of conscious awareness, as did their fear. Ultimately, our dependency on the natural world and our vulnerability is split off from awareness and, as Jenny expressed it, we "do just live a little bit in this kind of alternative reality that it's not climate crisis."

Disavowal and avoidance were also evident. On the one hand, participants acknowledge the relevance of the CEE, and therapists want to "make space" for a client's feelings about it, but at the same time neither therapist nor client opens up the topic. The therapists seemed to have a blind spot when it came to the CEE. They know it is there, but at the same time they do not see it. This turning away is a form of disavowal, a knowing about the seriousness of the CEE and at the same time making it insignificant in the context of the therapeutic encounter.

Another common defense was intellectualization, which serves to maintain an emotional distance from the subject. For instance, when I asked participants to tell me about their feelings about the CEE, they would often resort to a cognitive problem/solution response: greed is the problem, and connecting with Nature is the solution. Participants used the words "interesting" or "fascinating" as they reflected on the CEE, suggesting an observing self, curious but dispassionate. As Martin put it, "We touched on it when I was at university. I became quite fascinated with . . . the idea of the agricultural revolution, which we touched on at university, and anthropocentrism, man at the center. So I found this so fascinating, um, reading around it, and I've been reading about it since I've left university." Martin effectively maintains a distance from difficult emotions by maintaining a "scientist" position.

I have reflected on my own pursuit of a PhD as similarly defensive. By engaging with the subject matter through an academic process, I am effectively distanced from my emotional responses to it. In interviews, I found it tempting to engage in intellectual discussions with participants. I could pretend that this was purely academic, and any anxiety was deflected toward the challenge of gathering data and achieving my PhD.

Discussion

It would seem from the interviews I conducted that clients and therapists share a socially constructed view that therapy is for personal, inner-world issues. Additionally, beneath the surface, neither clients nor therapists felt safe enough to explore eco-anxiety deeply during therapy sessions. This speaks to me about two core concepts of therapeutic work: containment and the therapeutic third. In addition, a wider social perspective highlights a social defense against eco-anxiety at play.

Containment—the therapist's ability to receive powerful affect from the client and return it in a more manageable and less overwhelming form (Bion, 1962)—is necessary for the digestion, assimilation, and ultimately transformation of potentially damaging feelings into something the client can make use of. This "staying with the trouble" (Haraway, 2016) is a fundamental tenet of therapy. The difficult work of staying with painful feelings of helplessness, grief, despair, fear, and shame about the CEE could potentially lead to a reconnection with personal agency, a sense of love and commitment to the natural world, and motivation for engaging in positive actions aimed at reducing environmental damage or campaigning for change.

However, for this to occur, both therapist and client need to feel safe. The therapist needs to have explored their own emotional responses to the CEE and processed these in a safe space. They need to have experienced and survived the full range and intensity of their emotions. This then allows them to offer safe containment for their clients. Several of the therapists I interviewed seemed not to be able to offer the containment needed to explore the CEE with their clients, perhaps because they had not processed their own feelings about it sufficiently. Those who identified as eco-therapists were more likely to engage with it but still felt constrained by the therapeutic contract.

The concept of the therapeutic third derives from Winnicott's work. He proposed that in early infancy we learn the difference between what is "me" and "not-me" through the use of transitional objects that are both of me and not-me (Winnicott, 1971). This affords the developing infant an area of experiencing, a transitional space, to which both their inner life and the external world of reality contribute. It is also a space where play occurs, which is important because it allows for the accommodation to reality that can sometimes be difficult or painful. For the infant, playing with a reliable carer who is attuned to their emotional needs allows them to tolerate the anxiety experienced as they attempt to control their external world.

If we apply this to therapy of adults, therapy can be seen as a very sophisticated form of play, where both therapist and client are absorbed in a transitional space of creatively making sense of something the client finds difficult. Similar to the infant scenario, this encounter needs to occur within a reliable relationship where the client trusts the therapist to consistently attune to their emotional needs and contain any unbearable affect that may be experienced.

Benjamin (2004) describes this experience as "thirdness," a process of "letting go into being with [the other]" (p. 7). When this process of thirdness occurs, the therapist and the client co-create a shared vantage point that is outside each individual, leading to insights and new understanding. However, this is not an easy process to facilitate, and Benjamin (2004) contrasts thirdness with "twoness," where each person chooses between submitting to or resisting the other's perspective, a "doer and done to" dynamic (p. 9).

Benjamin applies this concept to therapy by considering the way in which a part-self of the client meets a part-self of the therapist (Benjamin, 2009). If the part-self being expressed by the client is not recognized and related to by an appropriate part-self of the therapist in such a way that they can engage in thirdness relating, then the transitional space is unavailable. This can all be out of awareness and need not be experienced necessarily as conflict. On the contrary, on the surface it can appear harmonious, a kind of pseudo-mutuality.

So how does this apply to my findings? Let us consider that when a client is making a side-mention relating in some way to the CEE, they are expressing their ecological part-self, the self that recognizes a kinship with and dependency on the natural world. If this ecological self is not recognized by the therapist because their own ecological self is defended, then no transitional space is available for creative exploration of the client's ecological self. If the therapist holds on to the therapeutic contract, sticking rigidly to the original focus of the therapy as a way of attending to their own needs—thus ensuring that the therapist's ecological self is not overwhelmed by the reality of the CEE—then the therapist has failed to be the reliable and trusted other whom the client's ecological self can depend on. The therapist has not been able to surrender to the process of thirdness, of meeting in the shared space. Therefore, the client's ecological self retreats, and the opportunity is lost. I believe this is what happens when clients' cues (side-mentions) about the CEE are not fully explored by the therapist. The therapeutic couple, in an unconscious intersubjective process, does not feel safe enough to contain the potentially overwhelming feelings experienced by both client and therapist.

Not only are we as individual therapists defended against our anxiety; so are our professional bodies, it seems. Social defenses against anxiety, first described by Menzies Lyth (1960) in relation to the work of nurses, operate through the adherence to professionally recognized practices and processes that serve to distance the practitioner from their feelings in relation to their work. For therapists, stressing the importance of therapeutic contracts and goals, and focusing on outcome measures and adherence to therapeutic models, although arguably good practice in many situations, can also serve to shelter the therapist from venturing into areas that may be potentially distressing for them.

The way in which the social, in this case the CEE, is kept out of therapy suggests a split of politics from therapy (Samuels, 2006) that drives our collective understanding of what therapy is for. Samuels (2006) draws attention to the way in which the external sociopolitical world is hardly ever mentioned in psychotherapy clinical texts, even though the founders of psychotherapy such as Freud, Jung, Perls, and Rogers saw themselves as social critics (Samuels, 2001). As the reality of the CEE continues to make itself felt, we are all likely to emerge from our climate bubble (Weintrobe, 2021) and experience significant shock. We are likely to feel vulnerable, angry, traumatized, shamed, afraid, and so on. This background collective dis-ease will be the context in which client and therapist are working.

Professional practice guidelines may then present those of us who are more aware of eco-anxiety with difficult questions. How do we ensure we are not "that kind of therapist" who imposes our own agenda onto the client and at the same time remain open for cues that the client may be ready to begin exploring their eco-anxiety? When and for whom is it therapeutically beneficial to hold the boundaries that keep the CEE out of the work? Conversely, when and for whom should the boundaries be softened to allow the CEE into the room? Working within each client's window of tolerance (Siegel, 1999) is crucial for safe practice. Additionally, moving between the inner and outer worlds of the client, disentangling the complexity of their distress, and developing an understanding of how inner and outer influence each other is key to a more permeable eco-psychosocial way of working (Rust, 2020).

Trainee therapists do not routinely receive theoretical or experiential training in how to work with eco-anxiety. As a result, it is not surprising that many of us do not recognize the cues that eco-anxiety may be present. We do not feel confident in knowing how to explore it and may even feel it does not belong in the therapy room. Although more is being provided in the form of CPD and by organizations such as the Climate Psychology

Alliance, training organizations have a responsibility to include such material in initial training programs.

Many of our frameworks derive from our individualistic culture, concerned as it is with the self, personal problems, and relationships. There is little scope in our current models for a collective lens. The CEE is a collective problem on a global scale; therefore, it presents a challenge to current ways of thinking and working (Bednarek, 2019a; Bednarek, 2019b). I am beginning to wonder whether working with the psychological and emotional effects of the CEE may be too big for a single therapist to adequately contain. Perhaps a different model is needed that is a better fit for the purpose. There are several examples of community-based models based on group working that have been developed to support people with emotional responses to the CEE, such as "The Work That Reconnects" (Macy & Young Brown, 1998), "Carbon Conversations" (Carbon Conversations, n.d.), and "Active Hope" (Macy & Johnstone, 2012). Such practices have been termed emotionally reflexive methodologies (Hamilton, 2019) and can facilitate the expression, containment, and processing of difficult and painful emotions associated with the CEE. They share the feature of being group based, creating a space for reflexive practice, and being held by a trained facilitator. Participants can gain a sense of resilience through interconnectedness with others and the opportunity to "link inner world with outer action" (Hamilton, 2019, p. 166).

Conclusions

The findings suggest that there is an urgent need to rethink the way we do therapy in relation to eco-anxiety. Therapists, clients, and our professional and social communities are collectively defended to varying degrees, often unconsciously, against the potentially overwhelming feelings associated with the CEE. As change agents, therapists have the potential to play a part in softening these defenses, leading to a more active engagement with the problem. However, we need support to do this. Our professional organizations need to provide adequate training that is both theoretical and experiential. As therapists, we need to be prepared to do our own therapeutic work around the CEE and to understand our relationship with it and our defenses. We need to learn to recognize the subtle cues from clients that may signal a preparedness to explore this collective problem and be willing to move into the transitional space where reality can be explored creatively, even when painful and difficult. More broadly, therapy needs to be re-visioned as a collective eco-psychosocial endeavor rather than a purely individualistic one. Perhaps we can only begin to talk about these feelings of existential proportions when we are in the company of others, when the container is big enough to hold us.

Frozen in Trauma on a Warming Planet

A Relational Reckoning with Climate Distress

Wendy Greenspun

This chapter starts with my sense of interconnection. As I write, a tapestry of voices and experiences resides in my unsettled consciousness: altered ecosystems in immense distress, communities suffering direct climate harms and environmental injustices, young people feeling the terror of a foreshortened future. Signs of what is crumbling on our warming planet. Infused with awareness of these layered perspectives, I will present the collaboration I have had with the person I will call Mr. R, who shares my despair about the state of the destabilizing ecosphere. I feel gratitude that he was willing to generously share his story and the story of our clinical work together. As Paulo Coelho (Goodreads, 2008) said, "The power of storytelling is exactly this: to bridge the gaps where everything else has crumbled."

This story represents only one facet of the multitudinous unfolding experience of the climate crisis. A single clinical case fails to represent the magnitude of the emergency—a reminder of the limitations of providing individual therapy when entire communities require support. However, as mental health needs expand in the face of increasing environmental disruptions, psychotherapy will continue to be an important branch on the growing tree of interventions, especially for those already struggling with mental health challenges (Doherty & Clayton, 2011). I hope that this clinical presentation helps to illustrate some of the complexities entailed in climate-informed therapy and encourages further conceptualizations of this important work.

Mr. R, a white, cisgendered man, came to therapy in a state of extreme anxiety in response to the climate and environmental emergency. He reported that while he typically handled adversity in his life with ease, climate change seemed beyond his ability to cope. After reading the dire Intergovernmental Panel on Climate Change report (IPCC, 2018) describing the short amount of time left to avert the worst consequences of the warming world, Mr. R went into an emotional tailspin. Like the rigid, repetitive play of traumatized children (Terr, 1990), he compulsively read every available piece of horrible climate news while ignoring or dismissing information about environmental activism or solutions. He described disrupted sleep and loss of appetite. He spoke with urgency about the catastrophe with anyone who would listen (and even with those who would not). He felt unable to calm his worries or take action. He said he felt frozen in terror.

People with climate distress and grief have begun to arrive on our clinical doorsteps. As the planet heats, as ecosystems destabilize, as the threat of social and ecological collapse looms large, the psychological fallout becomes evident. Emotional dysregulation may be the clarion call needed for each of us to take heed, breaking through the sense of complacency and complicity that keeps the emergency a lurking shadow in the background most of the time. Yet finding ways to bear the reality of this monumental problem is extremely challenging.

The climate crisis has been described as a "hyperobject"—Timothy Morton's (2013) word for a phenomenon so large in scale across space and time that it is hard to fully comprehend. It has also been described as a "wicked problem"—climate communicator George Marshall's (2014) phrase for difficulties that entail interlinking problems that defy simple solutions. From a psychological perspective, this existential threat represents manifest trauma of enormous scale, since the cascade of terrifying impacts shakes the very foundations of our belief in a secure attachment to the Earth.

And the trauma is not singular but infused with multiple levels of pain, a necessary expansion of the Eurocentric, individualistic definition of trauma (Craps, 2014) to a description that includes personal, relational, and communal experiences. Ecopsychologist Andy Fisher (2012) states: "All things are not simply connected; they imply or contain one another in their very being. . . . A philosophy of internal relations reveals a world in which there are no self-contained . . . entities but only fields of mutually informing relations" (p. 92). For those with a history of individual or transgenerational trauma, the climate crisis amplifies the impact of past wounds (Woodbury, 2019), further undermining a sense of safety, which now includes a precarious future. This becomes trauma within trauma. For those from frontline and marginalized communities suffering the most destructive climate and environmental impacts yet contributing the least to the problem, environmental injustice sits enfolded in the collective traumatic legacy of colonialism and systemic racism (Orange, 2017; Heglar, 2018; Ahmed, 2020). The climate crisis arises from the same toxic roots as these forms of othering, wrapped in the tendrils of greed, exploitation, and lack of responsibility for perpetrated abuses. This becomes trauma within trauma within trauma.

When faced with the enormity of these multiple traumas, we often turn away from the horrendous reality, understanding the truth on one level, yet simultaneously pushing aside what we know in defensive disavowal. Weintrobe (2020) says that disavowal protects our hearts from the moral conflict of knowing our complicity and shields us from the worst of our despair and grief but precludes the ability to mourn and repair. Even those of us willing to connect to awareness of the climate crisis may still split off the embedded systemic injustices. I realize as I write that I sit on the unceded tribal land of the Lenape people, just a mile from where the New York City slave market took place, facts I can easily set aside when I retreat to the pernicious oblivion of my white privilege. And the climate crisis itself results from the disavowal of painful truths: that life has limitations, that we are interdependent with the natural world, and that many aspects of human ingenuity are also leading us toward demise. On the intrapsychic, interpersonal, and larger systemic levels, disavowal can interfere with taking responsibility and finding avenues for essential engagement and action.

So how do we clinicians find ways to face this vast hyperobject without becoming overwhelmed or turning away ourselves? How do we conceptualize the responses of those suffering in ways that validate the reality of the crisis, recognize its interweaving with other psychosocial trauma, and understand the psychic meaning in terms of an individual's personal history without reducing climate distress and grief to a mere displacement (Haseley, 2019)? How do we begin to metabolize the pain that spans past, present, and future, affecting human and more-than-human alike? How do we reckon with our own climate distress, since we are entangled in the same planetary disruption as the clients we seek to help?

My own awareness of the climate crisis started like raindrops in a pond, splashes of recognition each time I read a news article about the catastrophic consequences of our warming world: the dramatic melting of arctic ice, devastating droughts, wildfires, hurricanes, and unprecedented species loss. Each moment of troubling recognition quickly dissipated into the normal rhythms of my life and my immediate personal concerns, since my social and geographic location protected me from firsthand impacts. Robert Jay Lifton (2017) says that when we continue to live in a functional environment, we find it hard to grasp existential threats like climate change.

My awareness grew to a steadier stream when evidence of the climate emergency touched me more directly. On a visit to the mountains in my home state of Colorado, I witnessed the marked devastation of forests, where the usual glorious blanket of green conifers had turned into a graveyard of arboreal skeletons. Increasing temperatures had prevented the usual winter die-off of pine beetles, and the proliferating insects were devouring the trees. The familiar solace I gained in returning to the alpine landscape of my childhood was replaced with a deep mourning for the life and beauty that was lost. Harold Searles (1960) noted that the nonhuman world can be as much a part of us as the parents who raise us—exerting influence, shaping parts of our inner experience. My sadness lodged as a dim presence in the background of my mind.

A flood of recognition was unleashed when I read a fictional depiction of a dystopian future, where the characters lamented that individuals and governments had known of increasing ecological devastation but failed to act. Stories have a way of breaking through the wall of psychic protection, helping us face the painful truths we usually push away. The veil of my defenses was fully lifted as I pictured my daughter and others of her generation admonishing our passivity as they faced traumatic stress. Waves of guilt, anger, sadness, and fear engulfed me. Hoggett and Randall (2018) describe an "epiphany" or sudden awakening to the realities of our warming world as an often-important initial step in the emotional trajectory of climate awareness.

The next step in this trajectory is the "immersion" phase: a period of talking, thinking, and acting on climate change. In my own process, I read vociferously on the topic as I worked with a trusted colleague to create and teach a psychoanalytic seminar course on the climate crisis. As we prepared the class, I read more pointedly about the overwhelming realities of ecological instability, the political and economic obstacles to making the urgently needed changes to our way of life, and the disproportionate impacts on marginalized communities. I began to feel increasingly distressed by the embedded layers of trauma in what I was learning, drowning in extended periods of sleeplessness, preoccupation, anger, sadness, and fear. Hoggett and Randall call this the "crisis" period in climate awareness, often marked by a sense of urgency, disillusionment, and intensely destabilizing emotions.

Many in the climate psychology field have noted the importance and even helpfulness of destabilizing emotions in the face of the environmental emergency, since disrupting the psychic and behavioral status quo can help us take in the urgency and find new ways forward. Bednarek (2021) calls this a "necessary derangement," while Kassouf (2022) pointedly states that the "cataclysmic realities of climate change call upon all of us to cultivate catastrophic thinking" (p. 60). Yet to make use of this crisis phase requires a capacity to process wrenching emotions and to bear seemingly unbearable pain, most easily facilitated in the presence of supportive others.

During my own crisis period, I turned to my coteacher for support and mutual processing of our grief and fear. We met regularly to mourn ecological losses, to rage at inequality, to acknowledge our terror and guilt. We used humor to manage overwhelm and shared unbridled tears. Over time I also joined climate activist and climate psychology groups for additional sources of solidarity and purpose. The poet Ryunosuke Satoro aptly stated, "Individually, we are one drop. Together, we are an ocean" (Goodreads, n.d., para. 1). In connecting with others, my internal seas quieted, and I felt better able to both cope and engage.

Hoggett and Randall call this final phase of the climate awareness trajectory "resolution"—a period marked by a greater sense of agency, more proportional emotional responses, and less negative preoccupation. Resolution may be akin to transformational resilience (Doppelt, 2016) and posttraumatic growth (Tedeschi & Calhoun, 1996), concepts that highlight how traumatic experiences can lead to new and stronger capacities. Resolution does not imply that painful reactions stop occurring; rather, it entails a greater ability to stay present with the feelings. This capacity becomes particularly important when embarking on clinical work with climate distress, as I learned with Mr. R.

From the start of our first session, I could feel my heart rate accelerating as a wave of anxiety rushed over me, setting off a cascade of emotions. As Mr. R described details of species loss and rising greenhouse gases as well as his vision of an apocalyptic nightmare for his daughter's future, I found my own sense of despair, that fire I could usually calm to tolerable embers, reignited like dry tinder. Despite all the time I had devoted to processing my own climate anguish, in the immediacy of Mr. R's pain, I joined his experience viscerally. The interpenetration of our psychic worlds pointed to our shared relational system in the face of this terrifying reality and to the ongoing countertransference challenges in doing climate-focused work. My own crisis feelings echoed within his, inhibiting my openness to try to understand him more fully.

By the second session, my internal wildfire was doused as my defenses quickly took over. Unlike our first encounter, this time I felt emotionally detached, as if I had moved from a resonant affective connection with Mr. R to a much more distant stance. My therapist role became a hardened, intellectualized shell, which pushed the anxiety and grief far away but seemed to interfere with my full presence in the session. I sheepishly noted to myself the relief I felt that Mr. R was now holding the distress and I could remain safely outside of it.

I was curious about the pendulum swing of my emotions and what it might mean about Mr. R and our burgeoning relationship. For this ongoing exploration to proceed, I knew I wanted to remain emotionally open to Mr. R's difficult feelings, to swim with him in those churning waters. As part of my resilience approach in doing climate-focused therapy, I strive to cultivate a "window of tolerance" (Siegel, 1999), the capacity to stay present with unsettling affect without either hyperarousal or defensive shutting down. In this spirit, throughout my work with Mr. R I utilized colleagues for support and paused to self-regulate, engaging in calming strategies such as meditation, deep breathing, and walks in Nature. These practices provided ongoing sources of grounding, strategies I also encouraged for Mr. R over time, and I was able to better sustain my clinical acumen.

As the treatment proceeded, we began to understand some of the personal underpinnings of Mr. R's climate distress. His family history revealed an unprocessed transgenerational trauma that easily mapped on to his response to the climate crisis. When Mr. R's mother was the same age as his teenage daughter currently, she was subjected to prolonged exploitation by an older man, with the explicit knowledge of her family and with no sources of safety and protection. She was never able to discuss this abuse in a meaningful way, sharing it only once briefly with her husband. In unpacking his responses to this horrible experience, Mr. R wept as he saw the numerous links between the harms that had been perpetrated on his mother and the environmental harms to Mother Earth.

Although Mr. R had been unaware of his mother's trauma history while he was growing up, he lived with the sequelae of what remained unspoken and unprocessed. Though often quite loving, she drank excessively, withdrew in the face of conflict, and isolated herself frequently, leaving Mr. R without guidance or comfort when he felt upset. Hopenwasser (2018) and Menakem (2019), among others, have spoken of the complex processes by which trauma gets encoded, embodied, and passed on to subsequent generations. From an attachment perspective, Salberg (2015) noted that the children of trauma survivors experience a "missing presence" when they encounter the parent's dissociative tendencies. This absence further perpetuates the impacts on the child, since the traumatized parent cannot help the child learn self-soothing or mentalization (Salberg, 2015). In one illustrative childhood memory, after Mr. R learned that cigarettes cause cancer, he felt terrified that his mother, a smoker, would get sick and die. He tearfully begged her to stop smoking, but rather than empathically responding to his fear and desire to protect her, his mother laughed and turned away. Mr. R recalled his sense of despair and a profound helplessness in the face of impending tragedy, an eerie parallel to what was now occurring in magnified form in the climate emergency.

I want to note here that, as with Mr. R, when a client's personal history seems to align directly with their reactions to the climate crisis itself, we may be tempted to consider the early trauma as the pertinent issue and the planetary emergency as only a symbolic representation. But discounting current material reality by privileging psychic reality belies the essential truth of their interwoven nature and, at its worst, mirrors a bystander's negligence in the face of actual harm. In fact, Woodbury (2019) describes climate change as a "superordinate form of trauma," encompassing and triggering individual, cultural, and intergenerational traumas. Climate-aware therapy asks that we hold the both/and perspective of complexity: climate distress is both a terrifying, psychically destabilizing threat and is filtered through the lens of an individual's history, strengths, and vulnerabilities.

Some of Mr. R's vulnerabilities related to his dearth of emotional coping strategies. While he had developed a capacity for certain forms of self-sufficiency in the absence of his mother's attunement, he had few resources for managing when he was suffering. His emotional dysregulation seemed like an ecosystem out of balance, where essential resources had been depleted and the diversity of inputs needed for healthy functioning were diminished. Trauma was his invasive species.

So what helps systems return to a state of balance and well-being? How could I facilitate ways for Mr. R to move through his terrible distress and find a more sustainable equilibrium? Zora Neale Hurston (2009) mused, "The present was an egg laid by the past, that had the future inside its shell" (p. 96). My work with Mr. R entailed a relational journey of "emotional composting" (Dunlop, 2015), where together we turned over multiple layers of individual, intergenerational, and collective history and embarked on a path to process his suffering.

We spent considerable time focused on his fears about his daughter, for here was the arena where past and future psychically collided in the present. Mr. R's belief in the inevitability of environmental collapse and his feeling powerless to protect his daughter from a devastating future encapsulated his mother's unmetabolized trauma and how she had not been shielded from harm. In addition, because he could not process or mentalize (Fonagy & Allison, 2014) his own distressed feelings, he was unable to help his daughter manage hers. The only protection he could imagine was walling her off from any awareness of climate change. He would hide newspaper articles on the subject, turn off television shows about global warming, and even hoped to block the Earth Day curriculum in his daughter's class. From Mr. R's perspective, he would hold all the pain of the climate emergency, and his daughter could remain blissfully unaware.

Randall (2005) and Lertzman (2013) have each described the process by which individuals who are trying to ward off despair may project their environmental concern onto those who engage in climate activism, who then serve as containers for the split-off collective environmental distress. What does it mean, then, to position oneself as the container and holder of pain for others, the role Mr. R was constructing with his daughter? I reflected on what I had not previously realized was an enactment of this same dynamic in our second session, when I had noticed my relief when Mr. R felt the upset and I could emotionally sequester on the sidelines. Mr. R observed that he played town crier in many of his relationships, sounding the alarm of climate doom for those who seemed less concerned.

We explored both the useful and the detrimental aspects of playing this role. Mr. R said that alerting others to the climate crisis felt quite urgent and essential, given the prevalence of those who minimized its significance. At the same time, his inability to metabolize the layers of his distress left him in a perpetual dysregulated state. As we explored additional unconscious and historical underpinnings of his responses, he had this insight: no one had acknowledged or prevented his mother's traumatic experience, so maybe if he held tightly to unmitigated alarm, he could ensure that harmful truths would have to be recognized. This important awareness brought him relief; in the words of James Baldwin (1962), "To accept one's past—one's history—is not the same thing as drowning in it; it is learning how to use it."

We also looked at Mr. R's tendency to isolate himself from others when feeling distressed, rather than seeking comfort or a shared sense of purpose; in this way, he mirrored his mother's coping style of emotional retreat. Our therapy became one forum where he could openly describe the depth of his reactions, reassured that I, too, cared deeply about the climate emergency but also seemed to have ways to manage that awareness. As his trust in me and in the therapy grew, he was able to let me witness and hold some of his complicated feelings as he mourned present and future losses, described the guilt of his collusion, and waded into the murky waters of helplessness and despair. Bednarek (2019) wisely reflects, "In order to take in the enormity of devastation that we have caused in the world, we need to know how to allow our hearts to break" (p. 15). As these emotions became more mentionable and understandable, he was finally able to start discussing climate change with his daughter.

Several situations in his life helped increase his emotional connections and reduced his sense of isolation, which began to thaw his frozen, helpless state of crisis. When the coronavirus pandemic hit, Mr. R felt immediate panic. He initially responded much as he did to the climate emergency: he focused solely on how to protect his own family, stocking up on months of food and supplies to feel prepared. Over time, he increasingly tuned in to those who faced greater COVID-19 exposure than he and his family did: those in low-income communities and essential workers who were unable to remain secluded and therefore risked their health for others. Mr. R began a practice of sending pizzas to frontline workers, expressing his gratitude and care. He reflected on how helping others seemed to soften his existential panic, and he felt less alone. This mirrors Doppelt's (2016) ideas for building resilience in the face of climate change: "When people experience acute or chronic toxic stresses, new social narratives that shift their field of focus to something greater than themselves . . . can provide invaluable sources of meaning, purpose, and hope" (p. 2).

In a second important experience, Mr. R encountered a young Indigenous climate activist who helped open his eyes further to the plight of others. This young woman described how sea level rise had caused her people's land to flood, destroying the natural habitat on which they depended for sustenance. Here was a community already suffering the trauma of the warming planet. Mr. R became aware of how little he had thought of current harms to already marginalized groups, since his climate terror had been focused only on fears for himself and his family. We began to discuss various aspects of disavowal and complicity in harms, including white supremacy and our shared white privilege, factors that had not previously registered fully in his mind or in our work. These realizations increased his desire to find ways to fight climate injustice.

The third significant situation was when his father, who had previously been diagnosed with cancer, took a sudden turn for the worse. With his father's poor prognosis, Mr. R drove across the country to be with him. He was able to take care of his father physically and emotionally—joking, sharing memories, and discussing regrets, pleasures, and feelings about dying. In our phone sessions during that time, Mr. R was able to further process this loss. In the aftermath of his father's death, we noted his growing capacity to face these difficult feelings, which helped facilitate such a deeply meaningful experience for and with his dad. Mr. R reflected that this transformative encounter became an opening, a tunnel through the impenetrable wall of past and future trauma, where the ability to connect and mourn brought in some light.

As I have discussed Mr. R's immense growth, it is worth noting my ongoing experience during the treatment as well. As presaged in our initial sessions, throughout the course of treatment I often rode the oscillations of Mr. R's emotional upheavals, since at times his distress and pessimism surfaced parallel feelings in me. In my own climate journey, I have vacillated between the extremes of hope and despair about a viable future, a not-uncommon dialectic tension when immersed in climate distress (Lewis et al., 2020). Mostly I sit in uncertainty, the reflective attitude that my training as a psychotherapist and psychoanalyst has enabled. This has meant that I often haven't known how to respond to Mr. R. Should I help him find ways to believe we can prevent the worst outcomes, directing him toward activism (Baudon & Jachens, 2021)? Should I encourage existential reckoning in the face of inevitable deep loss and possible social collapse? I know my role is not to provide answers but to find ways to stay with the difficulty of this wicked problem and wicked moment, right along with Mr. R.

So how do we all try to bear what is happening, grappling with the ghosts of trauma past, present, and future as we consider the climate crisis? How do we move away from forms of disavowal and othering that have harmed communities and ecosystems and that may be hurling us toward the end days? Kassouf (2022) has noted that a "traumatized sensibility" may allow us to keep awareness open, rather than continuing to turn away. In my own climate journey, I have found that the reckoning requires courage and support. I treasure the multiple ways I find meaning, connection, and sources of resilience, expanding my reach beyond the consulting room, including conversations with family and dear friends, facilitating workshops and Climate Cafés to provide forums for reflective sharing and resilience, participating in antiracist dialogues at my psychoanalytic institute, and teaching mental health clinicians to weave the climate crisis into their awareness. Engagement protects against futility and can plant seeds to grow an urgently needed "culture of care" (Weintrobe, 2021).

And of course connection with the natural world, healing the split of our disrupted attachment, is another crucial ingredient as we forge ahead. William Wordsworth tells us: "Come forth into the light of things, let Nature be your teacher" (Wordsworth & Coleridge, 2011, p. 106). I reflect on the extraordinary wisdom of the more-than-human world and wonder if we will listen (Kimmerer, 2013; Mitchell, 2020). I try to honor what Thich Nhat Hanh (1993) describes as "interbeing"—the way all living things constitute a mutually dependent system. I guide my clinical work with systemic eyes, focused on bringing individuals into closer contact with disavowed aspects of themselves, other people, and the broader web of life.

In this spirit, I want to return to where I started, acknowledging the voices and experiences that were only peripherally included in the story I have presented. While Mr. R and I have struggled with the "pretraumatic stress" (Kaplan, 2015) of those who are climate aware but not yet directly impacted by climate disruption, our anticipatory anxiety cannot be equated with the acute and posttraumatic stress responses of those suffering actual harm from extreme weather events, or the continuous traumatic stress (Eagle & Kaminer, 2013) resulting from environmental racism. Yet separating past, present, and future traumas obfuscates the important recognition of temporal interconnection necessary for our capacity to learn from previous mistakes, truly face current and ongoing devastations, and make choices based on the needs of those who will follow. Lifton (2017) speaks of the power of "prospective survivors" who use anticipatory distress to drive courageous action. Powerful social justice movements have emerged from outrage at systemic abuses. *Transformational wisdom* is a term I use to describe the trove of capacities and knowledge that have developed in communities that have suffered injustice and provide exemplary models of resilience (see Woods, 1998; Mitchell, 2020). The commingling of these various forms of climate distress and the generative capacities that can blossom from their seeds may in fact be what helps us move forward.

From a glorious mountaintop vista on a recent birthday, Mr. R gazed out across the vast landscape in a moment of deep reflection. He noted the magnificence of all that was present, all that was threatened. He spoke to the tree against which he leaned, tearfully apologizing for the thoughtless acts of degradation and other foibles of humanity. With a profound awareness of interconnection, he wept and wept until his grief became a deluge, frozen trauma unleashed. Resolution was beginning.

Chapter Three

The Long Shadow of Colonialism

Climate Change and Thirst

Shelot Masithi

Editor's note: This chapter is an amended version of a presentation the author gave at the conference "Six Months On from COP26: What Have We Learnt?" organized by the Association for Psychosocial Studies and the Climate Psychology Alliance.

I am a twenty-three-year-old psychologist and climate activist from South Africa. I grew up in three different villages: Tshandama, Dzimauli, and Tshitavha. Growing up in Dzimauli, I had the privilege of playing in the mountains, rivers, abundant waterfalls, and lakes. We did not worry about municipally supplied water because we had the river flowing near our homes. This is where I learned to swim. Seeing the river barely flowing today is nerve-racking.

Tshandama is the second village in which I spent half of my childhood. Growing up in this village, we experienced severe water shortages. Severity is an understatement for the water-shortage horror we went through from the years 2012 to 2017. The longer the crisis continued, the stronger my anxiety grew.

My mind was always occupied with water: "What if tomorrow there won't be water?" "Do we have enough buckets to conserve water when there isn't any?" In 2016, we experienced water shortages for a whole month. The water buckets were not enough, and the water we had reserved for such a crisis ran out mid-month. That month dragged on. We did not have enough money to buy water from those who had boreholes.

I remember one day, in the morning when I was ready for school, I thought, "Should I take a bucket to school so that I can fetch water on my way back from school—from the school taps or the river?" My high school had boreholes and rarely ran out of water. I thought that if I took a bucket to school, my classmates might bully me for it. The thought of other village children being anxious did not cross my mind. I stormed out of my room and ran to school, without the bucket. Studying was a nightmare, as the water was now a tenant in my head. At the time, I had little knowledge about the experiences of other villagers and countries concerning water access and climate change.

Amid this chaos, my geography teacher started teaching us about climate change. I still did not think that our water shortage had anything to do with climate change. He spoke of it as an event that would happen hundreds of years later. When I started reading about it, I realized that climate change was already affecting our present.

In June 2017, during a council meeting with community residents, my local chief decided to source water from the mountain. From October 2017 to October 2020, water problems were resolved. Then, in October 2020, there was not enough water from the mountain either. Seasons have dramatically changed; it does not rain as we usually expect.

Now that I understand the links between climate change and water shortage, I also understand the need for psychosocial interventions. We have separated climate change and social issues as merely scientific issues, ignoring the social and psychological effects they have on people.

I had learned about Ubuntu. When I look at climate change and water scarcity, I have seen in my community that people can come together to solve a problem with a common goal. The Ubuntu in my people. They wanted to have water to continue with their lives.

I think about psychology and that the psychosocial approach is a necessity for these problems. In the face of climate change and thirst, we need psychologists to look at these problems with holistic approaches that accommodate diverse communities and perspectives. Ubuntu is an integrated approach that recognizes the individual, the collective, and the planet. Ubuntu is a South African concept that recognizes the intersectionality between us and Nature. Climate change is a collective catastrophe, so the solutions need to be collective too.

As we are reflecting on COP26 in Glasgow (United Nations, n.d.), we need to understand the role of psychology in addressing climate change as well as the mental health effects of these problems on the collective. As a young activist, I can say that when we unite, we form a collective community with different branches, skills, and expertise, broadening our understanding of climate change, water scarcity, and what we need to do as mental health students and practitioners.

Global water scarcity is a systemic threat that four billion people face worldwide (Mekonnen & Hoekstra, 2016). Psychology in its individualistic form cannot help people in their diverse communities; it excludes them from the communities that make them who they are. At COP26, many world leaders and corporate leaders are united to harm us. They continue to approve coal projects that take away water access rights from the people (Olander, 2022). We cannot afford to discard our eco-anxiety.

The American Psychological Association (2022) agrees that the interconnectedness of climate change and mental health can no longer be ignored. Human beings burn out trying to solve these problems individually. The task to address climate change is beyond the abilities, skills, and knowledge of one person, one region, or one perspective—it is all of us. Within us, we each embody distinct strengths. And it is those strengths that hold solutions in them. When we try to alienate ourselves from the community, we risk the error of solving the functions of the crisis instead of the causes.

Despite our awareness of the crisis and its psychological effects, progress is far away. I know that most of you do not know what it feels like to be thirsty. I recently learnt that 75 percent of European consumers pre-rinse their dishes before they wash them (Hogan, 2021)—a privilege that we, in the global South, can only dream of.

The latest Intergovernmental Panel on Climate Change report projects that about seven hundred million people in Africa could be displaced due to drought. Instead of us looking at ourselves as intelligent living organisms that question everything, let us reflect on how we are shaping the problems and how others perceive them.

As mental health practitioners, students, and activists, where are we going wrong, and how does that affect the impact of our interventions? This is our opportunity to rethink, reinvent, and redesign with the collective in mind.

The Visibly Invisible Shadow

Decolonizing Work in Environmental Movements

Nontokozo Sabic and Malika Virah-Sawmy

The world has become a stage for the collective shadow. (Zweig & Abrams, 1991, p. xix)

Shadow, you who call yourself climate change,
I need to face you.
Shadow you are always there,
Always.
Faceless you are.
Omnipresent you are.
Visibly invisible you are.
Shadow, you who call yourself climate change,
Will you kiss me to death with your mouthless mouth?
Or will you make me love this planet like I have never loved?
If I could only see your eye, shadow,
I could hate my denial of you a bit.
But you're my wendigo shadow,
My own consumptive pathology.

—MALIKA VIRAH-SAWMY

In 2019, I (Malika) wrote a poem: "Shadow, You Who Call Yourself Climate Change." Even though I had studied sustainability at the University of Oxford, I was having difficulty being in relationship with climate change. Most of my education was about the physical mechanisms of this phenomenon, how scientists know such things, and how to deal with those who reject the science. I found climate change abstract, mechanistic, and faceless, and at the same time "visibly invisible," as I wrote in the poem. I worked as a climate activist and a sustainability professional but felt something pathological growing in me. The poem arose out of this context.

It was thanks to my contact with Nontokozo, my coauthor of this chapter, that I began to have words for the collective shadow I had felt but failed to articulate for so long. I began to see that climate change had been a visibly invisible forum for the collective shadow in its thwarting of our most well-meant intentions.

Before George Floyd, Greta Thunberg, and the social movements they unleashed, it would have been impossible for me to face the collective shadow of climate change. To be allowed to bring to the environmental movement the trauma of my ancestry—Indian bodies coerced to replace African slave labor in the colonies—was simply inconceivable. Yet now I can see that climate change is intimately linked to the culture of white supremacy and its related behaviors. This is the shadow that is in plain view.

Henderson (2019) says, "Climate change is the result of the so-called 'developed' societies becoming more socially and materially complex over time by rapaciously accumulating energy resources from so-called 'developing' areas of both the Global South and from within their own borders. . . . This was, and continues to be, a racial process, as whites accumulate/d wealth by violently dispossessing racialized others of their land and resources. . . . Climate change is an artefact of global industrialization via racial colonialism, with wealthy nations hoarding resources and disproportionately causing the climate emergency" (p. 987).

Robin DiAngelo (2016, p. 146) has this description of white supremacy: "White supremacy does not refer to individual white people per se and their individual intentions, but to a political-economic social system of domination. This system is based on the historical and current accumulation of structural power that privileges, centralizes, and elevates white people as a group."

The death of George Floyd in 2020 led to global protests and intensified the Black Lives Matter movement. His death was a tipping point that challenged colonizing nations to confront in ourselves, in our relationships, in our institutions, and in our systems the insidiousness of white supremacy. Greta Thunberg, another inspirational figurehead of a social movement, spoke truth to power and drew attention to the fact that "our house is on fire," while industrialized nations who are most responsible for the dangerous levels of CO_2 in the atmosphere carry on with business as usual. George Floyd and Greta Thunberg helped me to confront the collective shadow of climate change, a shadow I had felt for a long time but had been unable to voice.

Working on Our Denial of the Shadow with Ubuntu

Shadow-work is predicated on a confessional (and sometimes cathartic) act. (Zweig & Abrams, 1991, p. 240)

Nontokozo and I got to know each other through the Deep Adaptation Forum (DAF), an international online network fostering support and collaboration in relation to climate-related societal disruption and collapse (DAF, n.d.). Nontokozo was invited to work as a consultant for DAF, focusing on antiracism and decolonizing efforts.

Both Nontokozo and I grew up in a world full of incoherency. We were not able to make sense of our worlds. We were both mentored into an Indigenous world—that of Sangoma in South Africa for Nontokozo, and that of Indian-African fusion cosmologies in Mauritius, where I grew up. And we were both subjected to a colonial education with its colonial language, methods, and standards. *Neither* of us was taught how to bridge these very different worlds that we inhabited. Nontokozo says: "I am a Sangoma and I affirm it every day. I struggle with it every day, and I struggle with what I have been taught by Christianity. Living in a modern world, I have been educated in a Western way, and I have absorbed white behaviors. And I am also still walking in the world of ancestral Indigenous wisdom. White supremacy tells me that I am inferior. It is a painful journey to bridge the two worlds."

Nontokozo has offered decolonizing and diversity practices in the spirit of Ubuntu. The philosophy of Ubuntu derives from a Nguni word, meaning "the quality of being human." Ubuntu manifests through various human acts in social, political, and economic situations, as well as among community and family. It runs through the veins of all Africans and is embodied in the saying: *Umuntu ngumuntu ngabanye abantu* ("A person is a person through other people"). Ubuntu can be seen and felt in the spirit of participation, cooperation, co-creation, warmth, openness, and personal dignity.

"Ubuntu calls on you to stay open, to stay in humility, in love, in care, in visioning for a better world," Nontokozo tells me. She continues:

But to stay in Ubuntu, while helping white people face how their white supremacy impacts on all of our lives, comes at a cost. In the process, the disease becomes visible, you see the monster, the shadow that lurks in every corner of our daily lives.

It is important to recognize that white supremacy impacts on all of us. White bodies suffer and benefit from it, while people of color have internalized it. It expresses itself through the trauma of inferiority and having to assimilate or adapt, and through treating each other according to this norm—e.g., people speaking "good" English, or people of color with "white features" being considered more intelligent, beautiful, and deserving.

Decolonizing work is about holding space for the grief of how white supremacy moves in us. It requires courage and safe-enough spaces to talk about those wounds, and to reflect together on how our wounds are connected to white supremacy. It is also about bringing awareness to the ways people hold power in their wounded stories—who ranks highly and whose rank is low in the story that is being told. How are stories of white supremacy used to express identity? What is the wound in that need, and at whose expense is this need being met?

Decolonizing work rooted in Ubuntu means we can hold the disease of white supremacy, the monster, the shadow that also lives in environmental movements with love and compassion, and that we can do so collectively. Ubuntu reminds us that we are all related and that our futures are interconnected.

Nontokozo points out: "Ubuntu in decolonization practice is like a well, it's a resource. You need to know that the well is there. And you need to know how to find it and how to drink from it." The well of Indigeneity is not only about healing, wellness, or global solutions, but also about staying with the trouble of coloniality. According to Nontokozo, "People connect to the oppression they know. So drinking from the well of Ubuntu is not just about accessing water from a ready-made 'source' to soothe the oppression that one is experiencing. Rather, Ubuntu in decolonial practices is allowing a new sense of reality to emerge, tending our wounds together, learning, unlearning, and relearning how our wounds are connected to white supremacy."

This is possibly one of the most important challenges of our time. It is about breaking through our denials and remembering what was stolen from us in the making of white supremacy. The wounds of not belonging, of not feeling worthy of life, of feeling insignificant and the loneliness of separation, are wounds that emerge from the long shadow of white supremacy. This new awareness can help us remember that human well-being and healing are deeply connected to the well-being and healing of all living beings.

Sharing Our Denial of the Shadow

Our mental well-being is dependent on our capacity to face reality. We can only face reality by breaking through denial. (hooks, 2014, p. 16)

It is increasingly recognized that the climate and ecological movements are entangled with a paradigm that underpins the Anthropocene—that of white supremacy. People who are BIPOC (Black, Indigenous, and people of color) rightly complain that the climate movement is filled with collapse-aware people, while racism seems to be an afterthought. The movement mostly represents white people's voices, ideas, and concerns.

When the climate debate focuses on threats to the current means of sustenance and security in the West, it is frequently overlooked that the means of sustenance and security have not been equally distributed or even present for most of the world. Indeed, they are built upon centuries of injustices that have led to the collapse of societies in the global South. The global majority has experienced societal collapse in one shape or form already.

Racism, colonialism, and other forms of oppression are baked deeply into the global predicament of climate change. This means there is an urgent need for decolonizing work. For instance, eco-anxiety has become a mental health crisis in the Western world, an anticipation of its societal collapse. Many climate activists are less conscious that eco-anxiety comes with the fear of losing the privileges of Western civilization, while other peoples or civilizations have already experienced some form of collapse because of the history of coloniality.

But what does all of the above have to do with psychotherapy? Might it be argued that psychotherapy itself—or, rather, some of its theoretical and practical aspects—is entangled with white supremacy, a paradigm that underpins the Anthropocene?

Treating eco-anxiety as a problem of the individual—the way many mental health issues such as depression and addiction are treated—would be absurd. Every wound comes with its context. The decolonizing work Nontokozo offers in the Deep Adaptation Forum involves the holding of the eco-anxiety wound with Ubuntu, that is, with love, care, and courage. It also means bringing Ubuntu to facing the shadow of white supremacy and oppression that eco-anxiety is rooted in.

We live in an unfamiliar time, when our actions may determine the livability of this planet. Nontokozo tells me, "None of us can know what decoloniality looks like. We are living a journey into the unknown. The destination is unknown. The result is unknown."

Tending to the Politics of Denial

> *From racism to gender inequality to poverty, to soil and ocean degradation and political violence, we cannot fix the systemic issues with explicit, direct correctives—that is not where they live.... They inhabit the inflamed scars of previous generations, they grow in the wounds that never healed, they have submerged, and now it hurts everywhere. (Bateson, 2021, para. 9)*

We carry our collective shadow of white supremacy everywhere: in technology, in science, in the media, in spirituality, in religion, in community work, and everywhere else. Bringing in policies for diversity and equality alone is not sufficient. Experience shows that those who dare to confront the collective shadow of white supremacy openly often end up feeling marginalized. We may need to understand the mechanisms of denial in order to know how to even see the shadow. Professor Jem Bendell (2020), the founder of the DAF, uses the E-S-C-A-P-E model to explain some of the motivations we use to justify our attempts to escape from unavoidable aspects of our reality:

+ E = Entitlement
+ S = Surety
+ C = Control
+ A = Autonomy
+ P = Progress
+ E = Exceptionalism

For Bendell, these mechanisms support the Western ideologies of individualism, consumerism, materialism, and neoliberalism, and other traits of the political-economic and social system of domination that constitutes white supremacy. Facing the collective shadow while actively taking responsibility not to respond with similar E-S-C-A-P-E patterns is paramount in any decolonization practice.

In this article, we have tried to tell a story of tending to our wounds, the wounds of the world, and their intimate entanglement with white supremacy. But the story does not stop with us, and it will not stop until more of us share how our ancestors met: what white supremacy stole from all of us, how it separated us, and how we met the collective shadow together.

Decolonizing Psychotherapy

Hāweatea Holly Bryson

Our world is at a socio-political-eco-cultural turning point. This can be seen as the deconstruction of covert colonial paradigms of modern culture. A dismembering process both external and internal. What does this mean for the field of psychotherapy?

I propose we begin by distinguishing two lineages:

1. the lineage of psychotherapy itself, which developed through colonial and academic advancement; and

2. Indigenous lineages of therapeutic acts and healing, which are largely excluded from the psychotherapeutic profession and historically punished, omitted, and minimized.

Indigenous therapeutic perspectives have much to offer in response to pressing needs at this consequential moment in time. Rather than an extractive process of taking from Indigenous perspectives in order to heal our current predicaments, a decolonization of psychotherapy itself is required, which has the potential to restore, reunify, and re-Indigenize the collective.

Personal Context

My cultural identity is as an Australian Māori woman (Ngāi Tahu, Ngāti Māmoe, Waitaha). This is a bicultural predicament that places me between two worlds. A colleague referred to this as "my white sock and my brown sock." In order to maintain a seat at the table of the profession of psychotherapy, it is necessary to arrange my Indigenous worldview and to step into a cognitive space of borders and established Western theories and concepts.

My professional context is as a psychotherapist *and* a Māori healing practitioner. I work within Indigenous communities in Australia (Aboriginal and/or Torres Strait Islander), Aotearoa New Zealand (Māori), and

Hawai'i (Kānaka Māoli), as well as with non-Indigenous people who frequently describe the experience of a "cultural void": a thirst for meaningful rites or ceremonies. Many of my clients acknowledge that their upbringing in the dominant modern culture has come at a cost to their genealogical roots, which are often not given relevance in their professional or therapeutic spaces.

The Initiation of the Time

There is a great cost in the refusal to revisit foundational histories. Our colonial past aimed to eliminate or otherwise submit, indoctrinate, and assimilate Indigenous people into modern white culture. We are now confronted with interconnected global crises that pose existential threats to identity, freedom, and belonging. We face the collapse of many of the familiar structures and systems that uphold the colonial and patriarchal systems. These have torn us from our place of belonging within the natural world.

This book you are holding in your hands asks us to consider how we navigate this time of change. As a collective body of civilization, this unraveling holds great potential for the emergence of a new way that includes the past while expanding beyond it. I suggest that the field of psychotherapy is undergoing the same tensions in how to meet these crises that arise out of its own foundations. So what will it mean to decolonize psychotherapy?

> *Change rearranges the items on the table in new configurations. Transformation sweeps the table clean. (M. Ringwalt, personal communication, November 2016)*

Psychotherapy promises its clients transformation. This urges the question: can the profession of psychotherapy undergo its own transformation to meet the demand of our time? Thus far, psychotherapy has adjusted its concepts of inclusivity and raised cross-cultural awareness while staying within the Western paradigm.

Falling apart is terrifying, and it often leads to an attempt to grasp the certainty of known structures. Fundamental transformation often ignites deep-seated fears, as it affects every aspect of the profession: it confronts our methodologies, paradigms, identities, and ways of being. Modern culture tends to associate the experience with failure. However, from an Indigenous perspective, systemic collapse has always been a story of necessity. Furthermore, personal, familial, and systemic collapse are seen as stages of transformation and initiation. Across Indigenous cultures, these stages are spoken of at length through rich metaphors, mirrored by our natural kin. The French anthropologist Arnold Van Gennep (1909) described what he learned from old traditions and summarized pancultural rites of passage through the study of ethnographic records. He outlined three distinct phases:

1. *The severance or separation:* A letting go of our old identity, roles, reference points, or ways of being. This can be an ending, a death, or a falling away from familiar structures. This stage of severance emphasizes what is unsustained, out of balance, or separated from the whole.

2. *The liminal* (from the Latin word *limen,* meaning "threshold"): Being in a liminal state is to be in an unfamiliar and unknown place of becoming. A threshold is known as a point of departure or transition. There is a sense of irreversibility; we are unable to retreat from the new awareness. As Meyer and Land (2003) put it, "A threshold concept can be considered akin to a portal, opening up a new and previously inaccessible way of thinking about something. It represents a transformed way of understanding, or interpreting, or viewing something, without which [we] cannot progress" (p. 1).

   In the etymology of threshold is *thresh,* i.e., to *thrash,* like the beating of the grain in order to separate it from the chaff (Liberman, 2007). This is an innately uncomfortable and ambiguous stage. One moves between what was and what is yet to come. Across Indigenous worldviews, chaos and challenge are clarifying forces that the initiate faces. Bayo Akomolafe describes this in the opening conversation in this book when he says, "This darkness is productive, it's prolific. It's how to sit here that is the issue."

3. *The incorporation:* This is our return from this cataclysm. What is the gift we bring back from the passage or ordeal? This final task is to incorporate it, to bring new insights into the *corps,* the *physicality* of our body, our lives, and our community.

What this prepares me for is the acceptance that as we fall apart internally, we can better meet our human and more-than-human world.

Acknowledging the Field of Psychotherapy as a Lineage

Decolonization, which sets out to change the order of the world, is, obviously, a program of complete disorder. (Fanon, 1963, p. 36)

Lineages live through us, our beliefs, and our perspectives. These are likely to be unearthed and clarified by the liminal experience, including any cultural conditioning that is occluded or unconscious. Lineages are not only genealogical. There are lineages that ground us and that contribute to the formation of identity, such as those of our occupations and practices.

From an Indigenous perspective, the attempt to decolonize psychotherapy requires a conscious declaration of lineage. Our profession often assumes that our field is the apex of the evolution of the helping professions, without explicitly naming the colonial ground on which this assumption rests. Its advancement and authority went hand in hand with colonial advancement. Standardized psychological testing and the evolution of conditioning studies (Pavlov) coincided with the suppression and vilification of cultural healing practices.

The cultural and geographical lineage of psychotherapy was not acknowledged during my Western education. Therefore, listening to stories of other cultures was surprising for me, as in this instance: "Western psychology as a modern science was introduced to Korea through American missionaries and taught at private Christian institutions. In the beginning of the colonial period (1910–1945), modern psychiatry came in largely by two streams: the missionary institutions that adopted a humanistic stance; the other through the Japanese institutions heavily influenced by German descriptive psychiatry, which had strict restrictions regarding access to it by Koreans. In the 1950s the Western model of counseling was introduced by American educators, and the missionaries from the United States extensively expanded their involvement in Korean society" (Haeyoung, 2013, p. 2).

In contrast, the lineage of Indigenous therapeutic processes and healing acts, which I refer to as "Indigenous psyche-ology," is old and can be distinguished into preinvasion and postinvasion eras. The postinvasion era has been predominantly defined by colonial and religious paradigms. "It is essential to underline the point that we are looking at parallel cultural histories: that of the European scientific Enlightenment, and that of the Indigenous cultures, histories which, until recent generations, had no mental contact with each other's worldview or practices. What contact there was in the early days was through opaque glass, distorted by a kind of suspiciousness, and mediated partly through Christian conceptions and misconceptions" (Petchkovsky et al., 2003, in San Roque, 2012, p. 97).

The interconnected web of relationships and genealogy of Te Ao Māori (the Māori worldview) views modern culture as the *teina* (younger sibling) and Indigenous cultures as the *tuakana* (older sibling). Through this lens, the lineage of psychotherapy has Indigenous psyche-ology as its *tuakana*. This would require the younger sibling to decenter itself and acknowledge its belonging and contributions within a family lineage. This is why in acknowledging the lineage of psychotherapy we must acknowledge that the origins and foundations of the discipline are culturally specific. It would fare better if this was explicitly declared. In Indigenous cultures, acknowledgment of lineage is orienting; when lineage is made visible, safety increases. It communicates respect, implies sovereignty between lineages, and opens relationships toward synergy.

I often hear people say they do not have a culture. Modernity is a culture. I imagine that what we are each seeking is to uncover and examine our deep culture, with its transgenerational lineages and all that exists beneath the surface. We have reached a dissatisfaction with surface culture, which does not suffice when in the great unknown.

Cultural Dis-member-ment

Culture is a normative part of the self that is not explicitly taught but is assimilated through participation in it. Culture is the pool in which we are immersed. The etymological root of the word comes from the Latin verb *culturare,* meaning to cultivate the land, and *cultus,* meaning "homage to" or "cult." It refers to how we cultivate a place or a person through a specific paradigm of what is most highly valued.

From an Indigenous perspective, culture is always part of healing, and culture needs healing. "Culture is a living process. . . . Culture, spirit and identity are linked across time and place to country and kin. Healing occurs when these reconnections begin to be made" (Atkinson, 2002, p. 204).

With regard to cultivating the "ecosystem" of our psychotherapeutic field, we need to consider what has been deemed most valuable and thus tended or monopolized, and what has been culled, rejected, or neglected from the "ecosystem." How are we failing non-Western communities with a professional field that requires dialects of academia and data systems born of Western thought? "The methodology of psychological healing is very specific to the psychic culture within which the method is developed and the art applied. There may be essential elements of therapy, or a *therapeia,* that are universal; but, generally, psychotherapeutic practice has evolved in methodology and style in such a way that therapy reflects and meets the zeitgeist of the population within which psychotherapy survives—i.e., predominantly a Western scientific mind frame" (San Roque, 2012, p. 94).

If psychotherapy only addresses the need to decolonize its practice through cultural competency training, we barely touch the surface. While cultural competency is theoretical and based on intellectual understanding, cultural consultancy is living, dynamic, and in a direct relationship. Cultural consultancy is necessary for reconciliation and to advance toward eventual cross-cultural synthesis that would nurture true mutuality. To begin this process, we must explicitly recognize the prevalence of having "culled" Indigenous lineages of therapeutic processes and healing acts from our "established ecosystem."

New Zealand's Tohunga Suppression Act (1907–1962) called for the abolition of Māori spiritual and healing practices. The term *tohunga* refers to a crucial member and expert of the tribe: healer, priest, teacher, medicine person, traditional scientist, midwife, and so on. Māori healing and cultural therapies were punished until fifty years ago, with convictions of fraud, fines, and imprisonment for up to one year. This included anyone fraudulently claiming any knowledge or skill in "crafty science" (§ 261, 1908; Stephens, 2001, p. 456). Equivalents exist across many native nations. In Hawai'i, legislation outlawed and similarly persecuted Native Hawaiian healing practitioners during the period 1905–1978. The Revised Laws of Hawai'i, chapter 89, section 1077, outlaws Hawaiian Native healers, known as *kahuna*. Directly targeting mental, physical, and spiritual health in a people is an effective political strategy of devastation. There are three commonly established key purposes for the Suppression Act: (1) to counteract the consequent rise of *tohunga* untrained in Western medical techniques (Lange, 1999; Webster, 1979; Stephens, 2000); (2) to prompt Māori health reform and neutralize powerful Māori leaders; and (3) to allay Pākehā (European) fears about Māori attempts to claw back power and representation that had been lost to them over the previous decades of colonization (Stephens, 2001).

The reality that Indigenous therapeutic and healing ways were persecuted tells us that political and transformative power features in our healing modalities. According to Hawaiian Reverend Kalani Souza, decolonizing our psyche involves our present-day need to formally rescind the "doctrine of discovery" for what he calls the "doctrine of relationships" (personal communication, May 2020).

The doctrine of discovery is a concept in international law asserting that lands not occupied by Christians were considered vacant and could be legally claimed. The doctrine of discovery, still active in many countries, has been applied countless times by the United States Supreme Court and the courts of Australia, Canada, England, and New Zealand (Miller, 2017). In contrast, to restore the doctrine of relationships is to apply this between peoples and between people and Nature. Also of note is that Western legislation underpinned the case for discovery claims on the basis of land not being "properly used," with European and American law and culture deciding what "properly" means (ICT, 2020). When I hear language that speaks of "proper use," I experience an uncomfortable echo in the rules and regulations within our regulatory bodies of psychotherapy and the accepted pathways of formal training. Cultural assimilation strategies (the process by which a minority is expected to assume the values, behaviors, and beliefs of a dominant group) are reminiscent of our profession's rules about what constitutes psyche-therapies and healing.

Indigenous communities criticize Western mental health professions for disempowering them by perpetuating the same power-over structures, such as the white savior complex. An Aboriginal auntie once said to me, "The worst thing services give us is pity, a saturated poison that serves you and patronizes us. All these helpers rely on us to be sick. We're not set up to be well, because it's your identity and vocation that benefits, emotionally and financially" (anonymous, personal communication, October 2014). Helping professionals tend to freeze, even when confronted with more subtle versions of this response, as they do not know how to support our need for deep culture that is at the core of our world. This not-knowing is at the heart of the liminal stage and is what we are being asked to enter into more fully.

Language and Listening

Language is our landscape in the form of sound. Mnemonics of connectivity. Our natural world "can be likened to a book that will, on cue, release the traditions and knowledge associated with them" (Noon and De Napoli, 2022, p. 6). Dialect is sprung from the sounds of the landscape in which we live. Native languages deconstruct, rather than construct, our cognitions and identities. The survival of native language is inseparable from thousands of generations of traditional psyche-ology knowledge banks.

The suppression and punishment of native languages come at a great cost to humanity. We are lost to an intimate belonging when we no longer identify the name of our place, when they have been replaced with names of people. "Sometimes the name is no longer in the mouths of the people. What is it like to have the name taken from your mouth? To lose the name of the place that is your Home?" (J. Atkinson, personal communication, August 2022).

Language is a system of power and control. I heard another Aboriginal auntie say, "The mainstream, I now choose to call it the 'make-believe-stream' because it tells us white cis-males are the central voice of the world" (anonymous, personal communication, August 2020). Indigenous names were commonly changed to English names like "Jack." "Think about that amount of power, when your name cannot even be spoken" (U. T. Pahi, personal communication, September 2022). Language is wielded as a weapon, a great blockade, and a minimizer. Colonized thought permeates our education system, which primarily involves research and reason grounded in the dominant paradigm (Hall, 1976). In my experience, these histories continue in what is a pervasive demand that any contributions to our field need to be verbalized in Western constructs and academic language in order to be taken seriously.

Anthropology is considered a valid method for connecting with other lineages of psyche-ology. My experience of anthropology at university and in how it has interacted with revered elders in my communities has been the opposite of listening. Ways of listening are central teachings in Polynesian cultures. Psychotherapists are schooled to become certified "masters" in listening. Yet clinical foundations and our methodologies, however holistic, maintain a trained filter, which we believe is helpful.

Let me illustrate this with a story I have encountered more than once: a psychotherapist befriended a *tohunga/kahuna* (revered healer to the community) and asked to publish their teachings; the product reads like a validation of their own core premises and methodology. The authors' writing validated preexisting concepts and theories of psychotherapy while completely missing the doorway into felt paradigms in the landscape of much older healing ways.

In Polynesian schools of healing, healing reflects the healee's capacity to listen, and attention is paid to what they can take in at a given moment. *Wānanga* (healing and teaching spaces) are experienced differently by each person present, depending on which layer of knowledge one hears and connects with. With this premise, the therapist who recorded the elder has recorded themselves. Anthropology's body of knowledge reflects the student, not the culture. How did the student become the educator?

As another example, when the church describes Indigenous "sorcerers and healing spirits," it is really describing itself, in shadow.

Visions of Wellness

There is a Māori proverb that speaks about the amorphous liminality of time: *Ka mua, ka muri* ("Walking back to walk forward. We walk into the future to walk backward").

Mua both means "the past" and addresses whatever is in front of us. *Muri* means "the future" and addresses what is behind us. When we are aware of our past, our awareness informs what is in front of us. Similarly, we walk into the future facing backward, as we cannot see the "future" directly. The past is a knowing guide, as we are extensions of generations that came before us. This perspective is vital in understanding the repetition of colonial trauma and how it unfolds in the present.

Aboriginal therapist Paul Callaghan asks us to reflect upon our notion of wellness: "For me as an Aboriginal person, to be well I need to be well in mind, body, and spirit, but so does everybody around me; including non-Aboriginal people and Country. Dominant culture says a lack of symptoms" (P. Callaghan, personal communication, August 2022).

I was moved by this reality during Bush Circle, a dual-diagnosis bush-adventure-therapy program for Aboriginal young people that I led at Weave Youth and Community Services in New South Wales. Many of the young people in this program had never been into the Australian bush, having been raised in inner-city Sydney. Their grandparents' generation are the children of the "Stolen Generation," forcibly removed from homelands by government agencies and church missions. These young people struggled with substance use, mental health issues, and constant precariousness in their access to basic needs.

They emerged from the program with a sense of purpose and meaning. They developed a sense of inner security, reconnecting to Culture and Country (land), through mentorship with Aboriginal elders, and they found their ability to be of service. Their stories led to the program being a flagship case study for the International Union for Conservation of Nature World Parks Congress in fourteen countries.

Upon returning to the city, many of these young people felt empowered and became facilitators within family and peer groups, including leading Nature excursions for those struggling at home. The extension of one's personal growth into the family system was the deepest motivator for maintaining change. This was also what disempowered them. In reality, one cannot do well or be well within traditional society when family members are in pain or experience discrimination, addiction, trauma, or danger. This evidently showed that one cannot succeed or rise above those loved ones when others suffer. In Indigenous lineages of healing, family, community, and landscape exist as one body. The idea of an individual being contained by their own individual healing process is a temporary measure, serving only to resource them for the greater task of healing or tending to the collective space. This reminds me of the Māori proverb *he waka eke noa:* "a canoe we are all in without exception."

Individual and Collective Trauma

Can you heal from an open wound? (J. Atkinson, personal communication, August 2022)

Modes of control over Indigenous people, such as policing and incarceration, continue. In Australia, Aboriginal people make up 50 percent of the prison population but only 2.2 percent of the general population (Just Reinvest NSW, n.d.). In an elder's words, "Our displacement and dispossession now puts our babies in prisons, from age seven, eight, nine. . . . They are lost, they can't make 'sense.' I marvel that we aren't all stark raving mad, as there is no language to describe what our people have been through. . . . Trauma is embodied in different parts of us. We must engage in a deep truth-telling" (J. Atkinson, personal communication, August 2022).

With the wide spectrum of psychotherapeutic approaches, how can we declare our mental health profession adequate when our Indigenous communities are disproportionately incarcerated, medicated, or homeless? How can we have faith in its approaches when such ongoing cultural genocide and transgenerational trauma exists in all of our regions?

But we have arrived at a time when we can no longer ignore that all of us suffer from disconnection. We must begin with our own disconnection. "We need non-Aboriginal people to heal too. Your own healing and feeling are what Aboriginal people need" (P. Gordon, personal communication, December 2022).

What Does It Mean to Decolonize the Therapeutic Space?

I began my professional path as a psychotherapist as an intern at a juvenile prison in Colorado. One afternoon every member of the therapy team had entered isolation to offer support to a thirteen-year-old African American boy. My supervisor offered me the final attempt to get him to engage. He had been mute since an incident in the morning.

I was ushered into a bleak steel room with a toilet. The youth was slumped in the corner. The correctional staff wheeled a chair in. I pushed the chair away and trepidatiously sat on the floor with the toilet between us. I took in our world for several long minutes and eventually remarked on how bare and cold it was. He had shifted to observe me from the other corner. He began to tell me about his outburst, which he had deliberately done to isolate himself to prevent tomorrow's release. He had been notified of a new custody arrangement in his stepdad's care and expected to be severely beaten after release. After I conveyed this to his social worker, he gratefully remained detained, albeit not in isolation, while new arrangements were found.

The team asked about my approach as the youngest, most inexperienced therapist in the facility. Answering this question might have elicited a reference to clinical approaches. But what if it was not a technique? What if it was not a methodology? What if it was just the decolonizing of psychotherapy? Decolonizing the chair, the hierarchy, the silence, and the focus on the problem in the cell?

In their article "Decolonization Is Not a Metaphor," Tuck and Yang (2012) speak to the imperative for "land-back." In our effort to decolonize, the land must be given back to its "creation stories, not kept to colonization stories" (Tuck & Yang, 2012, p. 6). Indigenous epistemological, ontological, and cosmological relationships are land-based, so this is a physical undertaking. Without land, we are ghosts. "Land-back" involves giving the healing room back, as much as it is a giving back of one's voice, body sovereignty, medicine, plants, and access and rights to sacred, ceremonial, and burial places.

Māori elder and member of Parliament Samuel Dovers has described being "brutalized" by the education system, rather than being "guided" by it. He says: "We had to leave our *reo* [language], our *tikanga* [truth/protocols], at the door. . . . The impact of that was everlasting, this agenda of disempowerment" (Neilson, 2020). So this raises the question: How do traditions enter the psychotherapy room? How do we decolonize the therapeutic space?

What if a Latina client were to be encouraged to bring their ancestral bundle with them or to open the session instead of the therapist? What if a Māori elder was warmly welcomed to bring their concertina frame with photos of their *tūpuna* (ancestor) or could easily check in with you that there would be no food or tea in the room for the session? For Māori, in the distinction between sacred and casual space, sacred space requires the physical absence of food.

What is the role of ceremony in psychotherapy? Ceremony can be defined as "an intentional action (or set of actions) done in support of a clear intention, with a clear beginning, middle, and end, marking a significant time or transition" (Asmus & Bryson, 2020).

Giving back the therapy room challenges the therapist's fear of losing control of the space, the seat of direction, attribution of meaning, and/or prescriptions of wellness. The possibility of allowing tradition or ceremony to enter the healing space requires trust on behalf of both parties. Trust on behalf of the client, who needs authority to carry this with them, especially with generational trauma and introjected shame. Trust that it will not be commodified, romanticized, minimized, or reduced to a construct. And trust on behalf of the therapist, to enter unknown landscapes and sit in nuance. The level of trust needed here is a precursor for the very kind of healing relationship being called for by our current upheaval.

Online sessions, for example, have allowed me to be welcomed into clients' spaces and to be shown sacred objects, meaningful symbols, animals, family members, and kin.

When clients come into the therapeutic space, there are cultural aspects to enforce intentionality and safety that may not be known to a

Western therapist. Common to Indigenous protocol, there is the "host" and the "visitor." It is the home culture that leads, which means when a client comes to your designated space, they give way to your culture and customs (known as *kawa* in Māori). I suggest that this plays a role in the therapy room with bicultural clients. Modern culture may not be aware that the second stage of the protocol means that once the visitor is welcomed, the host needs to intentionally invite them to be free to include their ways of opening intentional space, personally meaningful ceremony, or traditional sequencing that might follow a deep release or completion, and/or make requests around hospitality or cultural or spiritual safety.

Cultural protocols allow for the reciprocal creation of designated space for sacred or therapeutic work. Mutuality is an age-old practice across the most diverse and disparate of cultures, language groups, and landscapes. Mutuality is made possible by such protocols.

Anthropocentrism

I want to reference a story from an international conference for practitioners of Nature-based therapy. My role as Indigenous network coordinator for this conference was to ensure representation. There was an air of excitement and hope for coming together across our domains of practice. In the middle of the conference, the Indigenous network connected. The sadness that poured in as we gathered was palpable, along with disconnected hearts. It was initially hard to vocalize the reaction of our bodies.

The exasperations I heard and relate to are: Why are we still spending hours presenting evidence-based research that Nature is found to be healing? Especially among practitioners who are immersed in this encounter with people and dedicated to this field? Why is Nature still under the microscope? How often do we reference Indigenous people in the past tense, or as anecdotal evidence, unconsciously?

The longer we sit to discuss facts and figures on the therapeutic value of Nature, the more we diminish what we are advocating for. It activates a protective polarity in me—a need to withdraw both the sacred and the ordinary from this debate. Our grossly overempowered mental constructs hijack our relationships. To illustrate this as a metaphor, in leading with research, the felt sense and colors of what we are discussing are lost, and I can no longer draw attention to what was here; it has since flown away.

In this regard, I observed two different experiences from the Indigenous representatives in the conference room. The first was energized by the evidence and the hope of formal recognition, almost as if their faith had been restored. The other group of people looked deflated, saddened, and shut down, almost as if their faith had been crushed. For this group, it was like finding a padlocked brass gate on the mountain path, and instead of decoding the numbers to continue, they would instead turn away and continue the convention with the birds.

These segregations appear in the most well-intentioned spaces. In the inclusion of the natural world as a cotherapist and of Nature being healing for us, we easily forget the two-way reciprocity needed between us and our places. We speak to "climate change" in theoretical language. We rephrase it as "climate emergency," and our own physical body enters a state of emergency. We maintain scientific language: "global warming," which then requires us to assess, confirm, and manage who is responsible.

Eventually, we call on animism. In Hawai'i, when the volcano was erupting through the highway, I heard people say "Pele [the guardian of the volcano] is angry with us," mostly said by those relocated to Hawai'i in search of something internal. In this, humans remain on the pedestal of importance, and I witness this to be where we often become stuck in shame or guilt.

In our natural world, our *tuakana* (older sibling) will experience everything first. When we find ourselves in the liminal and turn to Pele to teach us, we can receive her domain of understanding, movements, and ancient nuance directly as her *teina,* for we are indeed the younger sibling.

When the trees get sick, we will follow. When we poison our waters, our body's water will be next to hold that poison. Our whales have been disoriented and lost to the shore from sonar violence, like a TV in the womb of our home. Our native kin became homeless, and we too become home-less. When I speak in a direct relationship, I see humans as animals and animals as us. When we are *only* human and separate, to the expense of the more-than-human, we grieve our place in things we cannot describe. Through a personal relationship and attunement, we will experience direct information and guidance for what is necessary. These are our decolonizing feet.

I recently heard how even our sky is becoming colonized (Noon & De Napoli, 2022). As Bayo Akomolafe mentioned in the opening dialogue, this is "psychic gentrification. We colonize the darkness in order to let it be legible to what we already know."

In sitting with our disintegration as initiation, I *wetewete,* or "look into," the Māori word for "the in-between": *ki waenganui. Nui:* big, great. *Nga:* plural, many. *Wae:* leg or limb. *Wa:* across time, space, and place. *Ki:* to lean toward and connect. To be in the *waenganui* is to be with the big and the many limbs across time, space, and place, as entryways for transgenerational, transtemporal healing.

Songlines

I want to end with these lines, which were spoken in recognition of Indigenous Australians and their lineages at the World Council for Psychotherapy for their unique and local contributions to psychotherapy among Indigenous people:

> Let us pause for a second to contemplate. Over how many millennia have the people represented before you held to this line? Perhaps for two thousand generations. Does this experiential transmission count for nothing in your search for the evidence base of psychotherapy? Is this not time enough to establish an experiential evidence base, time enough to make mistakes and recover? Is this not time enough to learn how to help and heal the body/mind? Might this be time enough to maintain the cohesion of self/soul/family/country through all the ups and downs of civilization?
>
> Now, in our era of cultural breakdown, the answer may be yes or it may be no, but we may at least and at last acknowledge that the Ngangkari (Aboriginal healers) have been working to heal and hold the integrity of kin and *kurunpa* [spirit/psyche] since before Mohammed, before Jesus, before Gautama Buddha, Radha/Krishna, Abraham, Zoroaster, Ram/Sita, and before the maternal nurturant cultures of the Black Sea and Old Europe which, perhaps, formed the cradle of our European Caucasian cultures.
>
> Surely this is something worth recognizing? Surely it is this that the Sigmund Freud Award, given today, acknowledges? These seven people sitting before you carry in their bloodline the healing experience of the civilized world, and the experience of maintaining a civilized world, no matter what the violence and the devastation. (Speech by Craig San Roque at the Sigmund Freud Awards, 2011, in San Roque, 2012, p. 103)

Chapter Four

Unlearning Anthropocentrism and Decentering the Human

Anthropocentrism, Animism, and the Anthropocene

Decentering the Human in Psychology

Matthew Adams

> *As the shockwave of the Anthropocene affects all life on earth, questions about human relationships with(in) the larger-than-human community are urgent and provocative. (Harvey, 2019, pp. 79–80)*

The word *Anthropocene* refers to the idea that the Earth's geological record has been transformed by one species—the human. *Anthropos* is Greek for *human,* and *-cene* is a substantial geological time period within the current sixty-five-million-year-old Cenozoic era. The "shockwave" of the Anthropocene signifies the speed and scale with which many landscapes, species, and habitats are being lost and threatened by the impacts of human activity. The list is by now numbingly familiar: climate crisis and related consequences such as extreme weather, desertification, rising sea levels, ocean acidification; other forms of air, water, and earth pollution; species extinctions; and habitat and biodiversity loss, all intimately connected to unfolding psychological and social forms of distress. The fact that these changes are the result of (some of) the activities of just (some of) one species is, according to a professional consensus of interpretations of the geological record, most likely unmatched in the unfathomably deeper history of the entire planet.

For many, the Anthropocene is a problematic label because it crudely assigns responsibility to all of humanity, thereby downplaying the specific role of historical, economic, and political systems—such as colonialism, industrialism, capitalism, and extractivism—that benefit some at the expense of others. And it is these systems, rather than humans as a species, that continue to produce profoundly unequal and unjust outcomes within and across the species divide. While I am sympathetic to these critiques, for me the term works well enough to provoke those urgent questions about human relationships with(in) the larger-than-human community. The idea of the Anthropocene, stripped of any residual hubris, should provoke a *decentering* of the human. It can encourage us to learn to recognize interdependence with other species and more-than-human worlds, and to question what it means to be human. As the late environmental philosopher Val Plumwood (2007) warned, "We will go onward in a different mode of humanity, or not at all" (p. 1).

Anthropocentrism—literally meaning "human-centered"—is a belief system intimately associated with the Anthropocene. It is often considered a, if not *the*, dominant ideology at the heart of the ecological crisis, alongside colonialism, industrialism, and capitalism. Most importantly, it is a starting point for exploring and developing *alternative* narratives to anthropocentrism that are a necessary component of challenges to the intersecting causes of our current crisis. After a brief introduction to anthropocentrism, this chapter discusses how we are going about unlearning it in academic and professional practices and how we might further decenter the human. That discussion is followed by a discussion of animism as a potential alternative to anthropocentrism, grounded in everyday practices.

Anthropocentrism

A dominant anthropocentric worldview is a barrier to society (and humanity as a whole) reaching an ecologically sustainable future. (Washington et al., 2021, p. 285)

The belief that human beings are the central and/or most significant entities in the world is deeply rooted in various Western religious, scientific, and philosophical traditions, including modernism and humanism. Anthropocentrism is not exclusively or unequivocally "bad." In challenging religious orthodoxy, for example, humanism has often been associated with progressive politics, such as extending rights and care to other humans. It has been a useful belief system to the extent that it has played a part in the development of human social organization, technology, and culture, providing tangible rewards of security, comfort, and pleasure for some human societies some of the time. However, anthropocentrism is also considered to be at the heart of a profound split between humans and the rest of Nature, and therefore a central culprit in the cultural and psychological origins and maintenance of ecological crisis (Washington et al., 2021). At its core, anthropocentrism—especially as it combines with other prevalent ideologies—denies that nonhuman entities have intrinsic value, legitimizes a lack of ethical and moral obligation to such entities, and rationalizes their subordination on a planetary scale.

Many academic disciplines have questioned the presence of anthropocentrism in their own theoretical and methodological traditions (e.g. Hayward, 1997). This chapter will not rehearse these developments but will turn instead to the professional fields most likely occupied by the readership of this book: psychology, psychotherapy, and allied fields. Where anthropocentrism is found in the theory and practice of psychology, it excludes the significance of human relationships with nonhuman others and the more-than-human world more generally. Dominant traditions in developmental psychology and attachment, personality, and mental health have historically made little or no space for the significance of connections to place, animals, or Nature. Mainstream theories of social psychology have rarely considered the extent to which identity and belonging beyond human categories matter to people. Until recently, nonhuman entities were certainly not considered to be active contributors to interpersonal, group, or social dynamics. Similarly, clinical and therapeutic discourses making sense of experiences such as trauma and loss have rarely incorporated our relations with other species and places. Where other kinds of relationships are considered, such as in the development of pet-attachment or Nature-connectedness scales, or in evaluations of animal-assisted therapy, little concern has been evident for the interests of nonhuman others involved, or to envision the nonhuman side of those relationships (Shapiro, 2020).

Thankfully, anthropocentrism is increasingly being made visible, questioned, and reflected upon in numerous fields, including psychology and related disciplines. Recent studies in psychology explore how anthropocentric beliefs underpin patterns of thinking, emotional defenses, and tacit know-how connected with exploitative everyday practices involving other-than-humans. A good example is a flurry of recent work on the psychology of eating meat (e.g., Rothberger & Rosenfeld, 2021). Numerous studies have established the psychological processes involved in maintaining both a generalized love of and care for animals on the one hand and meat-eating practices on the other—a form of cognitive dissonance. Research has suggested that people involve themselves in a range of cognitive rationalizations and defense mechanisms that allow them to continue eating meat. These include minimizing our responsibility for suffering and perceiving the animals we eat as being highly dissimilar to humans, having less capacity to suffer, and/or being less intelligent.

This work highlights that anthropocentric patterns of thinking are not only psychological; they are also embedded in cultural and societal practices. These include the physical location and (in)visibility of farms and slaughterhouses; language, advertising, and branding discourses that objectify and commodify other beings; and cultural rituals and social norms. Vitally, anthropocentric beliefs are coded into everyday realities. Social psychologists Bastian and Loughnan argue that "when immorality reappears, people are not lost for ways to defend their psychological equanimity; they just seldom feel the need to" (2017, p. 287). Other research points to the cultural ubiquity of anthropocentric beliefs in multiple settings, including advertising, education, literature, science, newsprint, social media, and everyday conversations (e.g., Pedersen, 2021).

The idea of anthropocentrism as a historically embedded, psychologically adaptive, and culturally pervasive belief system can leave us with a sense of being "locked in" to it, with alternatives difficult to imagine. But of course, this is an overly reductive take. In everyday life, there are still countless examples where (human) people still find ways to be open to the rest of life and honor the living world, even if we also still routinely engage in behaviors and practices that do direct or indirect harm. There are also many and increasing attempts across art, literature, film, and the wider culture to develop nonanthropocentric forms of communication, representation, and storytelling. But we do not always have the language, norms, routines, and rituals that adequately frame these practices for us professionally, privately, or collectively. In this sense, there is significant value in exploring and developing alternatives to anthropocentrism that are imaginative, tentative, opaque perhaps, but grounded in ordinary and everyday experiences.

Actively Unlearning Anthropocentrism

The next step after acknowledging anthropocentrism: the active unlearning of it. (Rautio et al., 2017, p. 1381)

There are many signs that anthropocentrism is being not just acknowledged, but actively unlearned. Work in disciplines like anthropology and geography, and much feminist, posthuman, and resurgent Indigenous scholarship and activism is coming up with conceptual frameworks and alternative professional practices that challenge anthropocentrism in the context of the Anthropocene (for example, in forestry, microbiology, and zoology). An emphasis on the interconnection and interdependency of multiple species, including humans, has become increasingly common. In the social sciences and humanities, multispecies and more-than-human methodologies are emerging (see Van Dooren et al., 2016, for an overview). These approaches make efforts to give "voice" to nonhumans in the research process, while acknowledging their limitations. All share a desire to recognize the ways in which being human is profoundly embedded in relations with landscapes, plants, and/or animals (e.g., Lien & Pálsson, 2021).

What might an active and practical unlearning of anthropocentrism look like in a psychological context? In considering examples, it is perhaps helpful to move a little closer to professional practices more familiar to those of us working in psy-disciplines. Taylor et al.'s (2020) "Rescuing Me, Rescuing You" project explores companion-animal-inclusive domestic violence social work practice that is grounded in human–nonhuman animal entanglements. A dominant discourse in this context is that the discovery of animal abuse in families is important "only because it serves as a red flag for interhuman abuse" (p. 36). The authors argue instead for practices that center the experience of companion animals, "acknowledging that animals are sentient beings whose own experiences of physical and emotional abuse is worthy of attention" (p. 36). They reach this conclusion after conducting research that made animals affected by domestic violence visible, not just via human caretakers' accounts but as agents in their own right. For example, animals were present at interviews, and questions included a focus on the loving and restorative aspects of the relationship before, during, and after experiences of violence.

The authors make a strong case for companion-animal-inclusive domestic violence service delivery in social work practice, highlighting the significance of provision for companion animals in victims'/survivors' willingness to leave; how shared experiences of violence can intensify existing human–animal bonds; and, relatedly, the importance of providing support for companion animals and human victims as mutually beneficial. Practical challenges to a more animal-inclusive service include staff willingness to recognize and advocate for the importance of human–animal bonds; places and policies that can accommodate them; and providing appropriate care for traumatized animals. The project and the author's priorities are all about recognizing nonhuman animals as actors in their own right, attempting to take on the perspective and experience of the companion animal; emphasizing transspecies relationality and interdependency as the basis for mutual suffering and flourishing; and cultivating a more-than-human attentiveness and a sense of accountability and mutual obligation.

A growing number of studies in allied fields are making efforts to meaningfully incorporate other animals and other species as actors (or persons) in framing research questions, research design, and ethical considerations, and to consider implications for *mutual* care, development, and well-being across a range of settings. Asking these kinds of questions in specific settings requires us to seriously consider nonhuman interests and agency, or reciprocal relationality from a psychological perspective—what Gorman (2019) calls a "becoming therapeutic together" (p. 314).

Similar motivations underpin recent studies of cat cafés (Robinson, 2019) and equine therapy (Peralta & Fine, 2021) and my own research into sheep grazing conservation programs (Adams et al., 2023). All these studies ask if and how particular human–animal encounters benefit both, while also developing theoretical understanding of these spaces as fundamentally relational. That is to say, these studies emphasize the nature and quality of what goes on *between* humans and nonhuman animals, rather than taking an either/or approach to human and nonhuman animal well-being. Such accounts should also demand humility, recognizing limitations in our ability to perceive and infer nonhuman animal agency and subjectivity while asserting it.

Another interesting example, and one directly relevant to therapeutic practice, is the experimental program Serenity Park, based in the West Los Angeles Veterans Affairs Medical Center (see Siebert, 2016). The project brings together two groups of traumatized individuals—parrots that have been abused and abandoned, and former soldiers suffering from posttraumatic stress—to engage in mutualistic therapeutic practices explicitly designed to promote cross-species healing. Serenity Park is an animal sanctuary that takes in parrots in need of care, who are then nursed back to health by US veterans struggling with various trauma-related issues. Parrots are well known for being socially attuned and responsive animals, and accounts suggest deep and mutually healing bonds are regularly formed during this process. Bolman (2019) argues that a unique "mode of well-being arises between parrots and veterans," which he refers to as "becoming-well-together" (p. 309). This is a dynamic that temporarily decenters the human by extending personhood to parrots in terms of their capacity to suffer, connect, and heal, and by advocating for an ethics of care that is not exclusively human, but is fundamentally relational and mutual.

It promotes a *transspecies* understanding of trauma and healing. Trans-species psychology (e.g., Bradshaw, 2010) emphasizes human and animal interdependencies in terms of psychological experiences, while also acknowledging difference. This is a space that acknowledges that "the willingness for beings with very different histories . . . but shared experiences of suffering to open towards each other, a process which is the linchpin of so much therapeutic labour, points us, however hesitantly, in the direction of another world" (Bolman, 2019, p. 310).

An active unlearning of anthropocentrism seems central to the other world Bolman hints at here. We can see many other glimpses of this world across academic work, research practice, law, activism, art, and popular culture, where we are being asked to see the world and ourselves from the perspectives of nonhuman others, to acknowledge their interests, and to understand the depths of our entanglement with other species and other life on the planet. In the resurgence of animist worldviews, this entanglement moves to center stage.

Toward a "Grammar of Animacy"?

Maybe a grammar of animacy would lead to a whole new way of being in the world. (Kimmerer, 2017, p. 133)

As an idea, animism is hard to pin down. Broadly speaking, it refers to a belief system or worldview in which things that are human *and* other-than-human (creatures, entities, landscapes, objects) are alive, animated by something in common. In WEIRD cultures at least, is it only the human side of this equation that we take for granted? WEIRD is an acronym for people from Western, educated, industrialized, rich, and democratic societies, which academic research has tended to accept uncritically as a universal norm, despite only accounting for about 15 percent of the world's population (see Hruschka, 2018). To WEIRD audiences confronted with the idea of animism, it is perhaps hard to shake off the baggage of a century of imperial and colonial discourses imprinted in early anthropology. European and Eurocentric anthropologists largely present animism as a "savage condition of the mind that sees all things animate and inanimate on the same level of life, passion and reason" (Lang, 1899, p. 19), positioning it at an earlier stage in the evolution of religion that ranges from primitive belief to supposedly more advanced monotheism and then on to scientific rationalism. Such dismissiveness was part of the broader imperialist discourse in modernist psychology, anthropology, and science, which positioned other contemporary cultures (especially those being colonized) as further back on an evolutionary timeline, with Modernity representing the enlightened present. That is why it was also detectable in children, most famously in Piaget's child psychology, as part of a developmental stage we leave behind (or as Freud would have it, ontogeny recapitulates phylogeny).

Recent "new animism" movements firmly shake off this view and acknowledge a sense of indebtedness to Indigenous knowledges, in which complex animist worldviews are a central tenet. Historically "animism" (though not named as such) has been central to different forms of Indigenous knowledge. "New animism" is a way of thinking that can provide alternatives to anthropocentric understandings of the world. There are plenty of scholars in disciplines like anthropology and geography in and outside of these fields, along with many feminist, posthuman, and Indigenous scholars and activists, such as Vanessa Watts (2013), Zoe Todd (2015), and Deborah Bird-Rose (2017), to help extend and deepen our understanding of animism in the context of the Anthropocene. Religious studies professor and scholar of animism Graham Harvey describes animists as "people who recognize that the world is full of persons, only some of whom are human, and that life is always lived in relationship with others. Animism is lived out in various ways that are all about learning to act respectfully towards . . . other persons" (2006, p. 204). Animist understandings of what it means to be human are fundamentally nonanthropocentric. The properties of humanness, such as mind, agency, intention, and culture, are not understood to be exclusive to humans, nor is "personhood" simply extended to nonhumans. I find this aspect of animism the most difficult to grasp, but the human and more-than-human worlds are always "co-becoming," to borrow Country et al.'s term (2015), in lively interrelationship. Through these animate interpersonal, social, relational connections, personhood is understood as constantly being manifest through shared practices of perception, attention, and obligation.

Of course, we cannot simply concentrate hard and magic ourselves into an animist cosmovision. In that sense, WEIRD cultures and individuals are probably a long way off. We must also be extremely wary of holding up Indigenous communities as retainers of some form of pristine animism that the rest of us can appropriate. In terms of appropriation, a key problem here is assuming that the adoption of one or two attention-enhancing or Nature-connection activities is equivalent to a hard-won animist perspective built by Indigenous peoples over generations.

Relatedly, it is problematic to assume that such standpoints and practices can be extracted from ongoing, specific place-based relationships and held up as universal. Indigenous knowledge traditions of animism are not a romantic worldview intact "over there." Rather than homogenous, ossified "traditions," forms of Indigenous knowledge are plural and dynamic, evolving over hundreds or thousands of years, as adaptive responses to the environment and embodied in reciprocal relationships with place. This includes responses to loss, destruction, and change that have deep historical roots in Indigenous experiences of colonial power (Whyte, 2017). When it comes to assumptions about animism being somehow intact in Indigenous communities, it is also vital to acknowledge that specific experiences of the Anthropocene are an intensification of the environmental changes imposed on Indigenous peoples since colonialism. Animism may be struggling to survive, adapt, and develop in deep collective histories of loss. As Kyle Whyte puts it (2017, p. 155), "Indigenous peoples [have long] witnessed the away-migration of their non-human relatives." Whyte sees the *renewal* of Indigenous ecological knowledges, including animism, as a necessary basis for self-determination.

I could go on to describe different tenets of animism, but instead I want to consider briefly what it might mean in everyday practice in non-Indigenous contexts. Or as Stacey (2021) asks, how can "people with no claim to ownership of animist imaginaries find their way into them, play with them," and even "become transformed by them" (p. 98)? There are many possible roads. While Indigenous communities endure in some parts of Europe (e.g., the Sámi peoples inhabiting northern Norway, Sweden, Finland, and Russia), it is commonly claimed that European societies also have their own Indigenous roots in what we now call animist worldviews. While the absence of authoritative and agreed-upon traditions or texts is an obvious obstacle, efforts are made to reclaim and revive animist traditions in one's own Indigenous and geographical context. Inspiration might be sought in evidence of the status accorded to trees in early Irish law and lore, for example, according to which trees were almost equivalent to humans (Fields, 2020), or in the claim that early Celtic saints were commonly "distinguished by their special rapport with animals" (Serpell, 2010, p. 22). Contemporary forms of paganism combine these reimaginings with knowledge of Indigenous practices. Such hybrids are potentially subject to the issues of appropriation described above, among other difficulties (see Rountree, 2012, and Fisk, 2017, for excellent overviews of the complexities involved). That said, the deliberate, reflexive adoption of animist worldviews in neopaganism and related developments offer a path, if not a straightforward one, for actively unlearning anthropocentrism and embracing a coherent alternative. But if contemporary paganism is likely to remain a marginal(ized) everyday pursuit for the time being, where else might we turn?

Various efforts have been made to try to articulate common qualities of an animist sensibility that grows out of our experience of the world we live in, rather than directly borrowing from specific traditions, and that speaks to everyday psychological processes. Amongst other things, Kimmerer (2017) focuses on the vital importance of the language we use to describe the more-than-human world, encouraging us to move away from seeing Nature as a collection of noun objects and instead consider a grammar in which rivers, bays, and trees are verbs, involved in being. This is a "richly inhabited world" in which Nature is made up of subjects, rather than objects, a world "of birch people, bear people, rock people, beings we think of and therefore speak of as persons worthy of our respect, of inclusion in a peopled world" (Kimmerer, 2017, p. 133).

Westerlaken's (2021) account of a "relational care ethics" echoes a common emphasis on learning to pay attention with all our senses to what nonhuman others are saying. Her specific focus is mundane, everyday encounters with other species "that are continuously appearing right in front of us . . . [as] starting points for further imagination and knowledge generation" (2021, p. 522). She includes her own illustrations as ways of posing questions, such as "How can we design public squares as suitable resting places for pigeons?" and "Can we think of trees as interlinked networks that care for each other?" Though such questions are difficult to articulate, Westerlaken argues that everyday examples like this are important in "doing ethics, in practice, rather than defining ethics, in theory" (2021, p. 522).

Echoing our earlier emphasis on more-than-human methods, others point to cultivating arts of attentiveness to the ways in which the more-than-human world is alive, vibrant, intentional, and communicative (e.g., Van Dooren et al., 2016). Felice Wyndham describes this attentiveness as a form of "enhanced mindfulness" commonly found in Indigenous communities: "an extremely developed skill base of cognitive agility, of being able to put yourself into a viewpoint and perspective of many creatures or objects—rocks, water, clouds" (cited in Robbins, 2018, p. 6). More prosaically, recent psychological research highlights the importance of "noticing Nature" for developing a deeper sense of well-being and connectedness. It follows that greater attentiveness offers ground for developing deeper levels of care, obligation, and accountability. Heightened attentiveness is even argued to be the basis for developing a felt sense of shared becoming, predicated upon an awareness "that the relationships humans have with non-human entities are reciprocal and contextual rather than unidirectional and abstract, and that as these relationships progress each entity shapes the other in meaningful ways" (Reid & Rout, 2016, p. 429). Shaping each other—reciprocity—is considered an important element in unlearning anthropocentrism.

Perhaps, as some claim, a growing interest in animism reflects a wider positive transformation taking place, the shock of the Anthropocene acting as an invitation to fundamentally reimagine the interdependencies of human and more-than-human worlds. Pamela Gibson argues that "were we to have real empathy for 'the other' in the form of plants, mountains, animals, Indigenous humans, we [the main beneficiaries of WEIRD societies] would find it *impossible* to live as we do" (2019, p. 69). For me, claims that we should embrace animism, while recognizing Indigenous traditions, are ambitious and hopeful but can feel remote. Similarly, exhortations to learn new ways of seeing or thinking can feel difficult, inaccessible, even threatening. In this discussion, an active unlearning of anthropocentrism, and a tentative adoption of a grammar of animacy, are grounded in practices we can perhaps more readily recognize. Moments already or almost present are a basis for nurturing our imaginative capacities, however modestly. In this sense, animism can be viewed as an engagement of "practiced affection" that recognizes a more complicated sense of personhood in ourselves and nonhuman others, a localized experience of being in relationship and an act of accountability, mutual obligation, and care.

Conclusion: Decentering the Human in Psychology

The wider challenge for psychology as a discipline remains a foundational one: not decentering the human as such but decentering a certain understanding of the human as one who is independent, bounded, superior, and assumed to be exceptional in its abilities to solve problems, suffer, communicate, socialize, and become encultured. At this juncture, we need a psychology that vocally adds to calls to Indigenize and decolonize educational curricula, challenging entrenched, often Eurocentric and androcentric coverage of topics, citation practices, theories, and methods (Kessi et al., 2022). It requires us to attend to the role of animals and other species in our mutual development, in processes of healing, forming identities, and creating a sense of belonging, community, and culture. It also demands that we account for the lives of animals in the history and present methods of our disciplines, acknowledging our ethical obligations.

It is also vital that psychologists and allied professionals acknowledge that attempts to decenter the human, challenge anthropocentrism, and develop meaningful alternatives, animist or otherwise, are inherently political. At the same time, a global shift to protect and defend the "rights of Nature" or Earth rights is well underway. This movement fights for natural entities (rivers, mountains, species, ecosystems) to have their rights to exist, thrive, and regenerate enshrined in law, on a par with the rights afforded to people and corporations. Despite facing powerfully opposed alliances of extractivist industries, lobbying organizations, media control, and government bodies, many of these movements are resurgent and gaining wider support and traction.

Finally, challenging anthropocentrism in and outside of our disciplinary homes encourages us to be attentive, creative, and open-minded in how we orient ourselves in everyday life, embracing, however tentatively, a "grammar of animacy." It is of course no coincidence that an "active unlearning" of anthropocentrism and the storying of alternative imaginaries is happening in the context of unprecedented climate and environmental emergencies that also speak of a crisis of the imagination underlying Modernity (Ghosh, 2016). Donna Haraway (2016) writes that "a kind of dark bewitched commitment to the lure of progress lashes us to endless infernal alternatives, as if we had no other ways to reworld, reimagine, relive, and reconnect with each other, in multispecies well-being" (p. 51). In radically readdressing our place in the natural world, an active unlearning of anthropocentrism is vital for lighting up those other ways. The words of the late, remarkable environmental philosopher Val Plumwood are offered as a final reflection here:

"Help us reimagine the world in richer terms that will allow us to find ourselves in dialogue with and limited by other species' needs, other kinds of minds. I'm not going to try to tell you how to do it. There are many ways to do it. But I hope I have convinced you that this is not a dilettante project. The struggle to think differently, to remake our reductionist culture, is a basic survival project in our present context. I hope you will join it" (2009, p. 13).

Psychotherapy, Anthropocentrism, and the Family of Things

Rhys Price-Robertson, Mark Skelding, and Keith Tudor

Soul healers have been taking action toward healing Souls and the Earth since *Homo sapiens* emerged, and "outlier" scientists, psychologists, and psychotherapists, e.g., Carson (1962), Shepard (1969), and Bateson (1972), have long been urging wider attention to the interface between human and planetary well-being. Nevertheless, it was not until the early 1990s that Theodore Roszak's books put ecopsychology on the more mainstream psychological agenda (Roszak, 2001; Roszak et al., 1995). One item on this agenda is the discussion and critique of the extent to which psychological theories address the world's numerous ecological crises—including global warming, biodiversity loss, pollution, deforestation, and ocean acidification—and enable therapists and clients not only to reflect on this but also to act.

However, most psychotherapeutic modalities tend to focus on human-to-human relationships, thereby deemphasizing the more-than-human world. This article examines some of the theoretical entanglements of psychotherapy in the anthropocentric worldview that lies at the root of our current ecological crises. We agree with Haywood's (1997) perspective that "anthropocentrism is the mistake of giving exclusive or arbitrarily preferential consideration to human interests as opposed to the interests of other beings" (p. 51). Alongside most if not all environmental theorists and ecopsychologists, we see anthropocentrism, in its various manifestations, as one of the key drivers of anthropogenic environmental degradation.

Is psychotherapy anthropocentric? Some would agree that it is but would argue that this is not a problem, as therapy is primarily concerned with the *human* psyche. Our response, with Haywood, is that ultimately this is an ethical (or axiological) question for therapy itself, which also raises the question of the ontology (essence) of therapy—in effect, the definition and scope of therapy. Ever since Freud's (2002) *Civilization and*

Its Discontents, the scope of therapy has included the social, political, and material realms. Psychotherapists have long been interested in forces and entities other than the individual human psyche—social structures, political institutions, communication technologies, material resources—and it is not obvious why the nonhuman realm should be excluded from this.

In this article, we argue that psychotherapy generally fails to escape the trap of human exceptionalism that places human subjectivity at the center of all theorizing, and we offer ideas for moving toward a more ecocentric perspective. First, we suggest that psychotherapy could "expand self" by including the living world in theory and in the consulting room. Second, we argue that psychotherapy could "decenter self" by drawing on ideas and methods that position humans as but one member of a vast ecology, what poet Mary Oliver (1986) calls the "family of things" (p. 14).

Expanding Self: Acknowledging the Family of Things

If our civilization is to survive and thrive, we must foster shifts in our collective perspective away from being a primarily self-centered species, with demands that must be met and interests that must be served, and toward seeing ourselves as part of a wider natural system in which we have responsibilities toward the rest of life (Juniper, 2022).

Psychotherapy has much potential salience in our age of ecological crises. Despite this potential, psychotherapy literature, and indeed the practice of many therapists, tends to be heavily focused on relationships between humans, with more-than-human entities and forces constituting a backdrop (at best). As Bednarek (2019) observes: "Our therapeutic discourse focuses almost entirely on human-to-human relationships. We do not include the absence of relationship with the living world in our diagnostic thinking of developmental trauma, attachment patterns, personality adaptations, and mental health problems. Equally, our notion of community, relationship, and kinship usually stops at the threshold of our social network or our own species. It rarely includes our relationship to trees, rivers, mountains, salmon, bees, or water flowing through our bodies" (p. 25).

There are heartening signs that mainstream psychotherapy is gradually becoming more sensitized to the living world, such as the publication of this book and increasing contact between most modalities with ecopsychology and ecotherapy (e.g., Barrow & Marshall, 2023; Keys, 2013). There are also ideas and resources within existing psychotherapeutic traditions that are yet to be fully utilized in efforts toward mitigating ecological crises. A good example here would be the post-Cartesian conception of self in Gestalt therapy, which focuses on the interdependency of organism and environment. In this view, self is a process of contact at the boundary of the human organism and its social and natural environments.

Siegel (Garrison Institute, 2015) could not be more clear on the need to expand concepts of self: "If the self is defined as a singular noun, the planet is cooked." As an example of how the self may be expanded to include the more-than-human world, we turn our attention to illustrating how "self," in Gestalt theoretical terms (Perls et al., 1951), is continually expressed at the boundary between the organism and its environment. Below we illustrate an experience Mark had that emphasizes organismic connection to the living (and dying) natural world, along with some of the human actions that current ecological crises demand.

The Planet Has Your Back

Mark Skelding

When my father began his dying journey, I returned to my parents' home in a small Cornish village, far from my own in Aotearoa/New Zealand. I began a daily practice of walking up the lane to the cliffs, and around the rocky headland back to the old harbor. It was a stunning autumn, and the various daily tasks settled together into a singular rhythm of deepening relationship with place, mother, father, absence, longing, belonging, healing, seasons, dying, and more.

One bright, warm October morning, waking to these unresolved issues, feeling constricted, I took my familiar walk up toward the headland. It soon left the village and became very "Cornish coast": all heathers, birds, and breeze. Shades and shapes dissolved the gnawing of existential issues, and my being became very present.

It dawned on me that this moment, these sensations, and this "me-ness"—including "my" issue—were intimately related. Something loosened up inside me, and I paused for a moment. I was moved to lie back on the heather.

Lying there, looking up into the vast, puffin-swooping blue, I dissolved into the moment, and was profoundly received. A private wonder arose as I realized this flowing not so much as being between me and something else, or even between an inside and outside, but rather as a movement within a much wider container that involves all elements and contexts.

It was the strangest thing: one moment, there I was, lying back, gently held in the soft heather, waves splashing on the rocks below, feeling deeply held, steady, centered within that time of profound change.

Then, with no preamble, everything shifted. Suddenly, I found myself no longer looking up into the sky, but instead experienced myself physically held to the Earth above and behind me as my perspective changed, and I was looking down and down into deep space.

I was now glued to the Earth, sucked back against this solid presence, palpable, hard against my spine. I felt the rising edge of vertigo; a touch of fear, too. I felt very exposed and vulnerable. There's no word that describes an aural felt sense, as far as I know, but that is what happened. I was part of this.

I barely found meaning. It was something to do with whether I wanted to align or not. There were choices, and consequences, and the planet was indifferent. But, if I so chose, alignment offered possibilities for support.

Next, I felt/heard a voice saying, "The planet has your back." I relaxed completely, stretched my arms out wide, and allowed the wonderful ancient cliffs to have my back, holding me up and allowing me to experience my tiny point of awareness in the biggest context I have ever known.

This was a transformational moment that has enabled me to risk and stretch and dare myself to step up, even when I have felt unsure, alone, or vulnerable. The words whisper, "the planet has your back," and while that has sometimes meant encountering the limits of my knowledge, capacity, or courage (actually, those happen quite a lot!), it helps me check my motives, my alignment, and most of all my purpose.

That the planet has my back has encouraged, guided, comforted, and corrected me in a process that has taken me deeply into ecopsychology, social ecology, and psychologically informed activism (see Skelding, 2020). It invites renewal, participation, and trust, whether walking in the old forest beside my house, meeting with a client, doing public speaking, or writing to remind an eminent body or institution that "our house is on fire" (Thunberg, 2019).

✦ ✦ ✦

The above example demonstrates how psychotherapy might explore ways of being that are more attuned to the living world than is usual. Such attunement will be necessary if psychotherapy is to remain relevant in the coming decades of unfolding ecological emergency. Mark's experience also begins to demonstrate what the expansive conception of self may offer to psychotherapy more broadly. In the dominant psychotherapeutic paradigm, where self is understood as separate from the environment, his experience may easily be dismissed as an unusual personal experience on a clifftop, or worse, as a dissociation or even a stress-related psychotic break. However, when his inseparability from the Earth is acknowledged, and when his nonordinary experience on the clifftop is contextualized within a broader view, we can understand it as a moment of numinous contact between human and nonhuman.

However, even this expanded conception retains a bias toward the human side of the organism–environment coupling. Self is expanded, to be sure, but it is still the protagonist of a story being understood from the perspective of a single individual. Thus, in the following part of the chapter, we turn to the more radical approach of decentering self.

Decentering Self: Finding Our Place in the Family of Things

Whatungarongaro te tangata, toitū te whenua (The people fade from view, but the land remains).

—MĀORI *WHAKATAUKĪ* (PROVERB)

We have argued that, with a shift of emphasis toward a more expanded conception of self, psychotherapy can avoid the particular form of anthropocentrism that involves ignoring, or at least sidelining, the natural world. However, there is a subtler and deeper form of anthropocentrism that is not as easy to resolve, even for the post-Cartesian traditions. For example, the solution to Cartesian dualism that the Gestalt therapy tradition has settled on is to focus on the contact between organism and environment. While this solution is effective for many purposes, it nonetheless ensures that there is a human being sitting at the very center of all inquiry; the more-than-human world is relevant only to the extent that it makes contact with the human (Price-Robertson, 2020).

Where does this leave the more-than-human world? Where does it leave the black-tailed wallaby, nursing her joey, evading wild dogs, feasting on tussock grass, sheltering from the midday sun, living a life beyond the human gaze? Where does it leave the varied seasonal exchanges between cedar, fir, and hemlock, and, as Simard (2021) has demonstrated, a tree's favoring of its own offspring even while support-ing others of the same and other species? The simple answer is that it casts these family members to the peripheries, unrecognized as distinct entities, restricted to "the phenomenal silhouettes they present to the human gaze" (Harman, 2005, p. 35).

The anthropocentrism of psychotherapy risks confining even some of the most progressive forms of psychotherapy to what Næss (1972) calls "shallow ecology," which sees the natural world as a resource to be managed or used—even, for example, as a therapist, to provide a therapeutic intervention *for* their clients. Næss stressed that there is nothing inherently wrong with shallow ecology and that it is in fact important for some to carry on the work of managing the environment, with the ultimate aim of benefiting our own species. Nonetheless, Næss is famous for his idea of "deep ecology," founded on the idea that "the well-being and flourishing of human and nonhuman life on earth have value in themselves," and that "these values are independent of the use-fulness of the non-human world for human purposes" (Næss and Ses-sions, 1986, p. 1).

In reading the current psychotherapy literature, it is possible to garner the impression that the only possible solution to Cartesian dualism is an approach that focuses on the point of contact between human and envi-ronment. Numerous bodies of literature demonstrate this is not the case. These include:

+ Various ecological and dynamic systems theories, including chaos theory, the Gaia hypothesis, deep ecology, and Weber's (2017) "erotic ecology"

+ The "posthuman turn," including Braidotti (2013) and other authors who explicitly challenge what Haraway (1991) refers to as the "leaky distinction[s] between human and animal, organism and machine, and the physical and non-physical" (pp. 150–151)

+ Actor–network theorists, such as Latour (2005), who reconsider the nature of entities and the locations of agency by exploring the material linkages between both human and nonhuman actors, which interact and negotiate with one another in complex social–material networks

- ✦ Speculative realists and object-oriented ontologists (e.g., Harman, 2005), who conduct what Bogost (2012) calls "alien phenomenology," which "puts *things* at the center of being," stressing that "humans are elements, but not the sole elements, of philosophical interest" (p. 6)

How these bodies of literature can inform psychotherapy remains an open question. However, it is clear that the above theories soften identification with familiar perspectives, challenging one to perceive the world as it exists outside of human experience. One way in which we can explore how we think and feel *as* the natural world is through poetic inquiry (Prendergast, 2009), which offers a "fluid responsiveness" (Chidiac and Denham-Vaughan, 2007, p. 9) to the environment. We could even see this as a form of alien phenomenology (Bogost, 2012), which—while acknowledging that we can never know if and exactly how nonhuman or "alien" entities perceive the world—suggests that it is our very openness "to understand something about interobject perception" (p. 65) that can stretch our thinking beyond the horizons set by human self-interest.

Theory and Practice

We are suggesting that some of our familiar theory needs to be deconstructed and reconstructed if it is to be relevant to our living and changing world. To that end, we have suggested ways in which psychotherapeutic theory is entangled with the structures of thought that underpin these problems, as well as ways in which we might extend and expand our theories and application.

There are many therapeutic concepts and practices that could incorporate the more-than-human world, as Bednarek (2019) suggests. These bring rich opportunities for collaborative, intermodal, cross-sectoral, and transdisciplinary inquiry (Prescott et al., 2022). Might our intake and assessment practices involve exploration of a client's relationship with the trees, water, soil, and wildlife in their local environment? How can we include carbon dioxide, mycelium, ravens, and bushfires in our understanding of trauma? How would our conceptions of progress in therapy shift if we could recognize, with Plotkin (2008), that healthy psychological development appears to follow "a general principle of psyche" and that "the deeper we understand ourselves, the more of the world we identify with and, as a result, the wider our circle of identity" (pp. 359–360)?

We gesture toward a more radical refiguring of the relationship between humans and our nonhuman relatives. The clinical question here would appear to be: how can we encourage our clients to occupy a more humble position in the family of things?

People arrive at ecocentric perspectives in varied ways. For some, it will be a gradual process; for others, a decentering of self may occur suddenly in the form of a peak experience, as Mark's vignette illustrates. His experience enabled him to recognize a "psycho-ecological niche" (Plotkin, 2021) within a transpersonal and transspecies "psychosphere" (Skelding, 2020). This is the collective interiority of the "family of things" that reflects Ferrer's (2002) notice of "emergences of transpersonal being that can occur not only in the locus of an individual, but also in a relation-ship, a community, a collective identity, or a place" (p. 136).

Strategies aimed at shaking, disrupting, or dismantling the centering of human experience have long been practiced by shamans, healers, spir-itual teachers, and psychedelic guides. The current ecological crisis creates a pressing need for such strategies to be employed by psychological guides such as psychotherapists. Of course, we are not unaware that, at a time when the familiar in the environment is dying, we are challenging the relevance of "familiar" theory, and that some may experience a loss in that. We consider it important to acknowledge and grieve the loss—of familiar people, environ-ments, and theory—in order to cope with and/or embrace change.

No matter how we progress in the shift from an anthropocentric to an ecocentric perspective in and for therapy, it is important that with each step we strive to maintain an "I–Thou" relationship with the beings of the natural world, seeing them as willing and available participants within a "communion of subjects" (Berry, 2013, p. 252) rather than as therapeutic resources, interventions, or objects to be exploited. However, even this—which positions us (humans) first—needs a final deconstructive twist, for which we draw on Schmid's (2006) argument for a "Thou–I" relationship and "an epistemology of transcendence" (p. 240) in response to the chal-lenge of the other. From this perspective, it is important to conceive of healing as at once an individual, collective, and systemic process that builds on immanent, embedded, prior-to-human patterns and structures through which our world has emerged (Weber, 2016). In other words, it is import-ant that we acknowledge, over and over, our place in the family of things.

Psychotherapy as Sumbiography

Dissociation in the Anthropocene and Association in the Symbiocene

Glenn Albrecht

For people in advanced industrial and technological societies, the very foundations of their security and lifestyle are simultaneously causing their insecurity and the imminent collapse of the society within which they live. The collateral damage to those who still live in subsistence or semisubsistence parts of the world will be immense.

The cognitive dissonance involved in this living contradiction is enormous and must be addressed at multiple levels. Because I am a philosopher, with no background in professional therapy, my suggestions are tentative and need to be developed further to become more efficacious in a pandemic of mental health distress.

I have, however, created a psychoterratic (psyche-Earth) typology that consists of new and older terms created by myself and others, which hopefully contributes to a better understanding of the relationship between the psyche and the state of the Earth. The psychoterratic is the domain that studies and applies these new and old concepts.

Heal Thyself: A Sumbiography

At the level of the individual, there seems to be little that one can do to "fix" global environmental problems that are overwhelming. As individuals, we are born into a form of society with no say about its major features. It is now accurate to describe late capitalism as the Anthropocene, or period of total human dominance over all major biogeochemical systems affecting the Earth. It is not until we reflect on the path we have taken within such an epoch, particularly as young adults, that we can retrospectively evaluate the direction in life that we have taken or have been forced to take. Some people reflect in this way voluntarily;

however, others might need to engage in actions that trigger such evaluation.

In order to help reconstruct the path that an individual's life has taken with respect to the human–Nature relationship, I have created the concept of a "sumbiography."* In my book, *Earth Emotions: New Words for a New World* (2019), I used this concept to explain to the reader how it was that I had come to write a book about human emotions and Nature. I wanted a term for the summation of the crucial elements of a person's biography, human and nonhuman, that coalesced into a coherent worldview.† I expressed the idea in these terms:

> A "sumbiography" is the term I use to explain the cumulative influences on my life, from childhood to adulthood, that have culminated in my wanting to write this book about the relationship between humans, other forms of life, and nature. These influences include my immediate family and the writers and themes that have come to define my life and my ability to feel, first-hand, the emotional richness of contact with nature. The meaning and importance of the sum total of living together with "nature," people, and other beings is what a sumbiography attempts to describe and acknowledge. (Albrecht 2019, p. 14)

A sumbiography can be produced in many formats. As a written document in the form of a personal biography, remembering and writing down the seminal events and influences that pertain to Nature, life, and death are critical in constructing a sumbiography. However, in the twenty-first century, we have at our disposal a record of the life span that includes hundreds, if not thousands, of photographs and film or video segments that capture these important moments in life. At the conclusion of the sumbiography chapter I very tentatively suggested: "Finally, it is also my hope that the idea of a sumbiography will be useful to others who wish to understand their own response to the challenges faced in this century. Educators, for example, could use this way of revealing values and

---

* This term is derived from the Greek *sumbiosis* (companionship), *sumbion* (to live together), *sumbios* (living together), and, of course, Greek *bio* (life) and *graphy* (from the Greek *graphein*, to write).

† The term that is used in contemporary English to describe this state is "mindset."

emotions as a teaching aid for student evaluation of the past and their construction of visions of the future" (Albrecht, 2019, p. 25).

Sumbiography as Environmental Education and Personal Growth

Bearing in mind that the idea of using a sumbiography has only been
public since 2019, there have not been many applications of its potential for education, self-enlightenment, and healing or therapy. I have
written more nonacademic essays that try to capture its importance
in the development of Earth Emotions (Albrecht, 2020b); however,
there are some notable examples of its emergent use over the pandemic years.

One pioneering effort that was brutally interrupted by the COVID-19
pandemic was undertaken by Marc Delalonde in both Greece and Germany. For a sumbiography workshop in October 2019, he wrote about the
point and purpose of a sumbiography in the following terms:

> Do you remember places, animals, and plants with whom you have
> shared a special bond in the past? People and experiences that have
> influenced the way you see nature and relate to it? We want to read
> about them!
>
> We will be writing outside, so you can open up your senses to the nat-
> ural setting around you and bring back memories of when you felt a
> connection with nature. Pick one and start writing about it!
>
> The sumbiography workshop is an intimate moment, where ten
> people with different life experiences and perspectives gather to
> remember and share the story of their individual relationship with
> nature. It is a time of introspection and self-reflection, where you can
> free your positive and negative Earth Emotions. Writing about them
> and sharing them with the group is a powerful way to feel at peace,
> more connected to your environment and to yourself. (Delalonde,
> 2020, "Workshops" section, "Sumbiography" header)

Another example of the importance of recapturing and interpreting
a Nature memory of the past and writing it down was given to me by a
young friend who was reading *Earth Emotions*. While in the act of assimilating the idea of a sumbiography, he recalled duck hunting trips that he'd

experienced as a child with his father. While never fully comfortable with shooting ducks, he had not reconciled this positive bonding experience with his father and his own views on the death of innocent creatures. While the father modeled joy and excitement in the act of killing, the son was at the beginning of a pathway that culminated in feelings of grief and mourning about the death of nonhumans as an adult. He is now pursuing a new career in the bird conservation movement in the United States. Here is a sumbiography that took a person from duck hunting to duck protection and beyond. My friend now knows a great deal more about himself and his current worldview than ever before. He found this experience enlightening.

In an academic context, landscape architecture students were asked to compile their Nature experiences using memory, images, graphics, and creative art. Rebecca Krinke, professor and director of graduate studies in the Department of Landscape Architecture at the University of Minnesota, set the following exercise for her graduate students: "In this assignment, you can start with your very first memories, or even before you can remember, by talking to your family about what they might remember about you and your relationship to the outdoors. Our focus is your personal emotional experience with the earth" (quoted in Arndt, 2021, p. 5).

In response, the students produced a variety of visual accounts to form their own sumbiographies. Many used images together with text to account for their changing values and emotions over time toward Nature. They also highlighted the importance of key people in their lives, such as skipping stones with Dad or picking flower petals with a grandparent while others highlighted free playing or exploring in the wild with others their own age as hugely important. All the personal creative responses constructed a timeline of experiences that, for them, helped determine their cumulative outlook on life. While each student had a very personal trajectory, there are points of intersection that reveal pivotal moments in their environmental education and the construction of their personalities.

In addition to the keeping of Nature diaries in order to document the phenology and ecology of place, an ongoing sumbiography documents the intersection of personhood, other people, place, and nonhuman life. It is the sum total of the *sumbios,* or the art of "living together," and it can reveal the richness or impoverishment of such a state. The identification of impoverishment includes a range of negative Earth Emotions that can be expressed, for example, as ecophobia (Sobel, 1996) and biophobia (Kellert & Wilson, 1993). It is the source of these negative Earth Emotions that can be addressed by new forms of therapy.

Sumbiography as Therapy

It is clear that the socialization of some young people includes engagement with Earth-centered people (mentors) and Nature-rich locations that give them a strong sense of place and a commitment to leading their lives in accordance with that life-affirming ethos and ethic. It is most likely those who have this ethic who choose to become, for example, landscape architects, or to work in other professions that directly engage that orientation and use it to make a living. Some are forced to engage in this "vocation" as volunteers because the social milieu in which they find themselves is not conducive to paid employment in their preferred domain. There is often a mismatch between social structure and pathways for creative outlet of Earth-nurturing employment.

However, deep alienation from Nature occurs in a world where the "extinction of experience" (Pyle, 1993) and the associated emotional impact that I call *meuacide** (Albrecht, 2021), or the extinction of our emotions, overwhelm the very possibility of positive Earth engagement. Pyle, as an important pioneer in the study of Earth Emotions, argued: "To gain the solace of nature, we must connect deeply. Few ever do. . . . In the long run, this mass estrangement from things natural bodes ill for the care of the earth. If we are to forge new links to the land, we must resist the extinction of experience" (Pyle, 1993, p. 152).

If the extinction of experience also takes away our emotional responses to life, it is meuacide that must be countered, as well as biological extinction. Another way of thinking about this emotional alienation process is achieved by using psychoterratic concepts to identify pivotal moments in a person's sumbiography.

While solastalgia is a chronic condition tied to the lived experience of negative environmental change (Albrecht, 2005), *tierratrauma*[†] (Albrecht, 2012) is the acute condition of the immediate impact of negative

---

* From *meuə,* an Indo-European root for words such as "remove," "motion," and "emotion," plus *cide,* meaning "to kill."

† From *tierra,* the Earth, plus "trauma."

environment change. One example of such a reaction was provided by Preston McMahon, a student at the University of Minnesota. In a succinct but eloquently expressed passage, he described his first experience of the ocean as a young man: "Seeing the ocean for the first time could not have been any more disappointing. The water was dirty, polluted, and unlike anything my childhood had led me to believe the ocean was. I felt disgusted as I left, feeling as if I never even saw the ocean" (quoted in Arndt, 2021, pp. 44–45).

A historical example of tierratrauma can be found in an early settler's reaction to the loss of native landscapes in her immediate location. Australian settlers cleared land of its trees by a process called "ring-barking," in which an axe was used to cut a deep ring in the outer layer of the tree trunk so water and nutrients could not move through the phloem and xylem vessels. An ancient eucalyptus giant could be ring-barked in minutes and would take months to dry out and die. Fire would then convert the dead biomass to ash.

One early twentieth-century pioneer woman, Ida McAuley, was so moved by seeing the result of ring-barking that she wrote: "I found to my horror that the whole of the most beautiful bush . . . had been ringed. I turned & came away feeling physically sick at the ruin of a place I love. I think of it at night and it keeps me wakeful and again first thing in the morning & all day long" (quoted in Holmes et al., 2008, p. 13).

If compiling a sumbiography reveals unresolved trauma with respect to Nature or perhaps a huge absence of contact with Nature and wild things, it is possible that its use can help a person move from a state of dissociation from Nature to one of greater association with it.

In the case of tree tierratrauma, engagement with land restoration projects and tree planting is healing. In other examples where much-loved street trees are subjected to constant threat and removal by town governments, tierratrauma and solastalgia can be addressed by working with others to stop the threatening process (Albrecht, 2018). There is even the possibility of having intimate, healing relationships with individual trees via email[*] (Burin, 2018). It may also be the case that appreciating the sumbiography of a particular tree will help humans appreciate at greater depth what it means to be a part of life and its processes. The intersection

---

[*] The city of Melbourne in Australia "gave 70,000 trees email addresses so people could report on their condition. But instead, people are writing love letters, existential queries and sometimes just bad puns" (see Burin, 2018).

of arborgraphy* (see, for example, Beale, 2007; Pyle, 1993) and human sumbiography offers the possibility of tree intimacy as healing.

Building from the *sumbios* in sumbiography, we can create a new form of analysis I call *sumbioanalysis*. In order to tease out the biophilic from the biophobic, a start can be made on evaluating the level of attachment a person has to the natural environment.

A series of questions can be asked and answered about the levels of attachment and the amount of emotional literacy a person holds toward aspects of Nature and life. In contemporary society, this has become increasingly important as cognitive dissonance, generational environmental amnesia (Kahn, 1999; Kahn et al., 2009), Nature deficit disorder (Louv, 2008), and ecoagnosy (Albrecht, 2017) are all present as barriers to a constructive (good) relationship to Nature.

With respect to Indigenous people, on top of all of the issues highlighted above, there sits colonization as a layer affecting all others. Forced removal from traditional territory, loss of language, forced removal from families (stolen generations), and explicit racism make the creation of a sumbiography both complex and emotionally disturbing. When doing collaborative research with two colleagues, anthropologist Linda Connor and social psychologist Nick Higginbotham, in the Upper Hunter region of New South Wales, we interviewed an Indigenous man about his reaction to the sight of the hundreds of square miles of open-cut coal mines and power stations in what was once his "country":† "It is very depressing, it brings you down. . . . Even [Indigenous] people that don't have the traditional ties to the area . . . it still brings them down. It is pathetic just to drive along, they cannot stand that drive. We take different routes to travel down south just so we don't have to see all the holes, all the dirt . . . because it makes you wild" (quoted in Albrecht et al., 2007, p. S97).

To empathize with and understand what "it makes you wild" means, a non-Indigenous person would require detailed knowledge of the particular personal histories of people and place. As will be demonstrated below, healing "country" also involves healing people. In Australia, there are now many examples of Indigenous people returning to their lands,

---

* Recording the history of individual trees as a timeline of ecological and cultural narrative.

† "Country" in this context means the biocultural region that a tribe/clan traditionally occupied.

repairing damaged places and the lives of damaged people. Their stories are instructive.

Healing through Earth Repair

In past publications on the issue of solastalgia, I have suggested a number of actions that might help the various forms of desolation being experienced by people and communities. In my first publication on solastalgia, in 2005, I gave the example of Indigenous rangers engaged in the repair of their own country.

A wonderful Indigenous woman, the late Cherry Wulumirr Daniels, a senior ranger with the Ugul Mangi women rangers from Ngukurr in southern Arnhem Land, explained how working with others to remove invasive weeds and trees had positive psychoterratic benefits.* Through her, I was able to get an insight into the emotional benefits of the active repair of damaged environments. As reported by Young (2004), the Daluk (women rangers) carry out environmental and psychocultural restoration. As Daniels put it:

> What the weeds have been doing in our country, in Australia, damaging, a lot of things, especially in our waterways, taking up much of the soil, taking up much of the water, so we have been looking at a lot of trees that are not ours, so we told our people to get rid of those trees, trees and weeds, and even feral animals. . . . And I'm very glad, it makes me feel happy inside when I see them do those things. Without me they identify an ant from our natural ant to the other ant that comes from out of Australia, like the Singaporean ant, the big-headed ones, they can identify which. (quoted in Young, 2004, para. 29)

It is these actions that restore positive psychoterratic states such as *endemophilia,* or love of what is endemic to your part of the world. Indigenous people now have occupation of their land and responsibility for its management in many parts of "remote" Australia. On country, in a reversal of colonialist dominance and overseeing, they can revive language,

---

* It needs to be understood that for Indigenous Australians, invasive plants and animals (often including colonizing humans) have the effect of destroying their endemic sense of place. The value of traditional knowledge is lost if invasives make their dreaming and songlines no longer functional. As a result, Indigenous ethics often comes into conflict with Western environmental ethics and traditions associated with animal liberation.

educate their children with traditional knowledge, and use affiliated cultural practices to reinforce their sense of place. As Tyson Yunkaporta (2019) simply put it, "Country is becoming well. You are Country. You are becoming well" (p. 260).

Repair of psyches damaged by damaged landscapes is now a global-scale possibility for all humans. Never before has there been so much work to do! Many before me have suggested that a dose of Nature in the form of immersion in wild or cultivated landscapes is critical for the restoration of the health of mindscapes (Pretty, 2017; Louv, 2011). In addition, getting to know place and its nonhuman inhabitants (education) is another way that individuals can restore their own mental health while restoring their home environment.

A sumbiography can reveal just what kind of emotional compass we have with respect to our personal relationship to this planet that supports life. I describe this relationship as *sumbiosic,* meaning "those cumulative types of active and purposive relationships and attributes created by humans that enhance mutual interdependence and mutual benefit for all living beings, so as to conserve and maximize a state of unity-in-diversity" (Albrecht, 2019, p. 200).

Transference

In exploring the emotional compass, a "sumbiotherapist" might be able to offer a pathway for a person to move from negative psychoterratic states into positive ones, from not living together (antisumbiosic) to living together (sumbiosic).

This form of healing has affinities with older, Freudian-inspired forms of psychotherapy, in which a distressed person can be helped by a creative engagement between a therapist and the "patient." In the contemporary context, "the patient" is likely to be suffering or distressed because of anxiety about climate chaos, the state of the world, and its future. The role of the therapist is to find elements of positive Earth connections in the past and help the individual use those connections to rebuild that dissociated material into a new worldview, where that valued facet of Nature is amplified and reinforced.

In conjunction with a meme (cultural replicator) such as the Symbiocene (Albrecht, 2019; Albrecht, 2020a), a sumbiotherapist can help an individual with the *transference* of that positive facet in the past to a future state, where it becomes associated and reinforced with a family of like-minded emotions. A new motive or motivation enables a process of transference where the isolated emotional moment can join or rejoin a coherent whole that is associative and "healthy." Once inside that newly opened space of good possibilities, further healing can take place.

An example of such therapy in action comes from Trebbe Johnson's *Radical Joy for Hard Times* (2018). In this book, she documents the restorative power of paying close attention to "wounded places," deliberately intervening in them, and performing a healing ceremony in the form of a symbolic work of art constructed from the very material that constitutes the wound. Johnson (2018) describes this process as an "Earth Exchange" (p. 151). In performing this simple act of symbolic healing, mental health is also addressed, in that this new motivation to repair what seems to be irreparable overcomes what otherwise leads to solastalgia and ecoparalysis (Lertzman, 2008). If this simple act of repairing a wounded place is accomplished, it opens up the possibility of wider acts of repair and reconstruction.

The more people engage with the damaged Earth, the greater the exchange between people and place, the greater the restorative power of this form of active therapy becomes. What seems impossible, such as removing the ghost nets (mountains of plastic pollution and discarded fishing gear) left in places like the shores of remote beaches in northern Australia, becomes acts of creation as the rubbish is removed and turned into works of art (Le Roux, 2016).

As such restoration expands, creative connections to other Earth Emotions become possible. Gradually, destructive psychoterratic experiences are sublimated or diverted into good or creative forms of expression. Terranascient (Earth-creative) emotions are liberated into new directions (e.g., music, art, and literature). That is how the Symbiocene, as a new era where *all* human attributes are associated, will be built as an act of human creation. In *Earth Emotions,* I stated the case this way: "Thus, in the global order, there is a parallel with the associative and dissociative qualities present in psycho-social good and evil. Ethics, in the symbiomental context, is concerned with the linking of associative psychosocial attitudes and institutional design with the patterns of symbiotically derived order to be found within life systems. These organic connections may occur in personal reflection, social movements, and ultimately institutions that are self-determined, self-organized, and self-perpetuating forms of symbiotic organic order. Allow these elements free rein, and you get the Symbio-cene" (Albrecht, 2019, p. 153).

Conclusion

In healing thyself, there is the potential to heal with others. As these con-nections grow symbiotically, there is a reversal of the dysbiosis (unhealthy life) that has taken only a short time in human history to build. Humans are social animals, so working with others to defeat dysbiosis is critical for a future symbiotic Symbiocene.

The role of therapists in shifting individuals and groups from damaged and dissociated states to healthy and associated ones is increasingly crit-ical. As the world gets further out of balance and crazy (*koyaanisqatsi**), those who can help individuals rediscover their foundational, life-affirming selves will be doing good work. Those teachers and healers who can put before distressed humans a positive vision of a good future will also be ethical healers for the Symbiocene.

In the Symbiocene, healing thyself becomes healing for the whole. There will be no need for psychotherapy to heal the human–Nature relationship. Once we are in the Symbiocene, we will never want to leave home.

---

* A Hopi Indian word meaning a world in chaos, a crazy life; a life out of balance.

Chapter Five

Collective Trauma, Fragmentation, and Social Collapse

Climate Change, Fragmentation, and Collective Trauma

Bridging the Divided Stories We Live By

Steffi Bednarek

Climate change reminds us that we are vulnerable and mortal beings who are utterly entangled with the environment we live in. It emphasizes the fragility of life and highlights that Western societal constructs are not as solid as we may think. The capitalist story traditionally promised safe lives into old age, control over the forces of Nature, and the idea of continual progress. But technical advances are currently unable to stop the laws of Nature from unfolding. We have entered the territory of literal and symbolic death, including the death of the familiar. The vastness of the losses already unfolding opens up a collective trauma field that will have far-reaching consequences for the mental health of many; in turn, the collective psychological response to adversity will shape the future of the world. We live in important times.

We are set for disruptive levels of global warming and may already have passed an irreversible tipping point (IPCC, 2018). No place on Earth will be spared the consequences, but those who already suffer from social inequality, poverty, and marginalization will be disproportionally affected. Climate injustice and the competition over scarcer resources are likely to widen the social gaps that already exist. Social unrest and mass migration are likely to soar (World Bank, 2018).

In order to avoid catastrophic runaway climate change, Western industrial nations will have to dramatically change their way of life. No nation is currently on course to meet the target of CO_2 emissions needed to keep global heating to the minimum of 1.5°C set out in the Paris Agreement. In fact, global emissions are rising rather than decreasing. Once the seriousness and scale of the problem sinks into public awareness, the risk of a global mental health crisis is high. How do we meet the enormity of these times?

For decades the scientific community assumed that logic and reason would inevitably lead to a logical and reasonable response. This has not been the case in the fifty-plus years that industrialized nations have known about the risks of climate change while increasing their carbon footprint. The failure to acknowledge the complexity of the human response has come at a high cost. There is clearly a discrepancy between logical thinking and the fallible human response to threat.

I argue that efforts to meet the challenges of climate change need to go beyond a mere reduction in CO_2 emissions. They require the maturing of the collective culture into a much larger capacity to process painful experiences while holding the interconnected, nonlinear complexity of life. This includes the ability to acknowledge fragility, to bear the unbearable with dignity, and to bring integration into the frozen and fragmented states of collective trauma.

Psychological analyses of the climate crisis have diagnosed a state of melancholia (Lertzman, 2015), resistance, disavowal (Weintrobe, 2013), and denial in Western culture. I argue that we also need to look through a collective trauma lens in order to understand the level of dissociation and inaction that we continue to witness. This is not as an alternative to other theories but in multilayered addition to them.

I propose that the phenomenon of personal and cultural fragmentation, explored here through the lenses of brain hemisphere balance, collective trauma, and the Jungian (Jung & Jaffe, 1995) concept of necessary suffering, is a relevant perspective on adaptation to a changing environment. I suggest that there is a need for collective ways to reclaim fragmented parts, bridge existing polarities, and meet the challenges ahead with maturity.

Divided Stories

In order to understand the lack of mobilization in the face of danger, it may be useful to take a closer look at the narratives that inform dominant Western values. The structures within which we live require a certain disposition in the general population. Capitalism in its neoliberal form has become hegemonic in Western culture. It has become a way of life that manifests in day-to-day experience. In the global North, capitalism has become part of our relationships with each other, ourselves, and the environment we depend upon. It has traded the idea of community for individualism, prioritizes profit over its consequences, and relies on unequal distribution of resources.

The Capitalist Institute (Confino, 2015) reports a direct link between the capitalist worldview and global challenges such as climate change and political instability, and it recommends that an urgent shift toward systems-based values is needed. This has enormous consequences, as capitalist ideals are woven into the fabric of Western culture. Many psychological theories reflect the capitalist values of individualism, materialism, anthropocentrism, and progress, and the concept of mental health itself can be regarded as the capacity to function symptom-free within a capitalist system (Bednarek, 2018).

Capra (1982) points out that the dysfunction of complex systems on the world stage is primarily a crisis of perception, where seemingly innocent collective everyday beliefs contribute to the stuckness of much larger, complex systems, and where the stories we tell about the world and ourselves serve as the connective tissue that holds things in their rigid place. Capra and Luisi (2014) stress that many solutions to global problems come from the same linear thinking that created them, whereas Nature is highly nonlinear, acting in feedback loops and forming and reforming wholes.

The philosopher Zygmunt Bauman (2000) writes about Western culture as being in a state of "liquid Modernity"—a state characterized by chaotic, ungovernable situations, where a reductionist focus on one specific problem in isolation is no longer adequate. The major problems of our time are complex, interconnected, and systemic and they need systemic and interconnected solutions that view life in terms of relationships.

The exponential growth of the technological and virtual aspects of life moves in the opposite direction. The project to eradicate vulnerability and mortality from human experience is not a mere fiction anymore. Artificial intelligence research talks about the development of the "transhuman" or "posthuman" generation, in which people are no longer dependent on their environment and are "enhanced" by technology, with the view that human and machine will converge beyond flesh-and-blood intelligence in the future. This suggests a vision for humanity in which fragility is engineered out of existence. Life is measured in years, not in connection, relationships, and Soulful engagement.

This positivist outlook relies purely on reason, analysis, and linear thinking. In an effort to understand the world, things are divided into their component parts. From this perspective, the world is inert, measurable, and controllable, and it can be understood through intellectual engagement and acted upon in a mechanistic fashion.

A holistic view sees the world in flux, and it values the embodied, sensual, spiritual, poetic, and relational aspects of life. This perspective allows multiplicities and contradictions, including the terrain of Soul or psyche (Hillman, 1995), which ascribes meaning to adversity and sees the world as an interconnected living system of ecological interdependencies (Bateson, 1972; Capra & Luisi, 2014).

Depending on where one's affiliations lie, personal ethics, values, actions, and solutions to individual and global problems will look dramatically different. From the positivist perspective, climate change can be assimilated in terms of facts and figures. Solutions are likely to be practical and technical, and communication relies on sharing information. A Soulful perspective, on the other hand, calls for an aesthetic engagement and a changed relationship with the Earth and each other, one where we become involved, entangled, and accountable.

Relational and mechanistic outlooks can either be viewed as mutually exclusive or as linked on a spectrum that connects one to the other. Bateson (1972) suggests that we need to move fluidly between these polarities, zooming in to a narrowly focused, targeted attention, then zooming out again to the interconnected relationships of the whole. This fluidity requires connection between the two brain hemispheres that govern these two types of attention. The brain's divided structure can easily lose the connection between the hemispheres and can fragment into fixed, compartmentalized positions. This is frequently the case in trauma responses, as I will explore later. Without the ability to shift perspective, there is a risk of "othering" those who do not fit within one's worldview, treating them as adversaries. This polarization into fixed positions seems to be on the increase at a time when we need an ability to change, work collectively, and think creatively.

In the following, I will discuss the phenomenon of fragmentation and compartmentalization from the perspective of brain hemisphere balance and collective trauma.

The Divided Brain

Each of the two brain hemispheres has important characteristics to offer. The left hemisphere deals in abstractions and categories upon which predictions can be based. It perceives things as fixed and known. This kind of attention enables us to examine, analyze, decontextualize, and generalize. It gives us control and power over a world that is seen as dead matter. The right hemisphere, on the other hand, is relational in nature, understands implicit meaning, and applies contextual thinking. It relates in an aesthetic and embodied way to the world and sees life from complex, changing, evolving, interconnected perspectives.

Psychiatrist Iain McGilchrist (2010a) illustrates how brain function is intrinsically linked to the dominant norms and values of society. McGilchrist (2010b) points out that in the history of Western culture, there was a fluctuation between times of hemisphere equilibrium and periods of left hemisphere dominance. In the seventeenth century the Enlightenment paved the way for the Industrial Revolution. From that point onward McGilchrist sees a shift into left hemisphere domination, intensifying over time, with only a brief attempt to rebalance the equilibrium during the Renaissance.

He describes a hypothetical culture in which the left hemisphere has won absolute control and power over the perception of reality. This culture is one where the living world is objectified and used at will. Narrow-focused attention leads to an increase in specialization, technicalization, and bureaucratization. Knowledge is more and more abstracted from experience. Holistic thinking is suspect. The world becomes more virtualized and modeled on mechanical ideals. The impersonal replaces the personal, and the focus on material things increases. Relationships are qualified by exploitation rather than cooperation. Governments become preoccupied with security, surveillance, and control. Compassion is replaced by rationality. Religion and a sense of wonder seem illogical and suspect. There is a difficulty in understanding nonexplicit meaning, and a marked desire for literalness prevails. A utilitarian approach to life dominates mainstream thinking and culture (McGilchrist, 2010b, pp. 70–71).

This largely describes the world we currently live in, where the gifts of the right hemisphere have become silenced and denigrated. The cost to the living world is evident in the predicament of our times. The solution is not a shift to right-brain dominance. Both hemispheres, with their distinctive versions of the world, have an important role to play; but their current relationship is far from symmetrical. McGilchrist (2010b) points out that "the trouble is that the left hemisphere's far simpler world is self-consistent, because all the complexity has been sheared off—and this makes the left hemisphere prone to believe it knows everything, when it absolutely does not: it remains ignorant of all that is most important" (p. 68).

We have witnessed the danger of extreme compartmentalization in the banal-seeming administration and bureaucratization of the Holocaust. Sereny's (1974) interviews with prominent Nazi leaders led her to conclude that it was systematic compartmentalization that allowed individuals to carry out genocide and still live with themselves. The Holocaust was orchestrated through bureaucracy, task division, and carefully abstracted language, where people became "cargo" and genocide was "a solution of the Jewish Question." The function of the right hemisphere was systematically excluded. This extreme compartmentalization allowed ordinary human beings, not monsters, to carry out monstrous acts as part of a day's work. The horrific became normalized through small steps into the unthinkable.

Gretton (2019) compares the psychological desensitization and fragmentation processes that enabled the Holocaust with contemporary contexts, suggesting that similar dynamics may be at work in corporate executives, who carry out systematic ecocide (Higgins, 2010) and are prepared to let people and ecosystems die if it increases profit margins. These "desk killers," as Gretton (2019) calls them, transgress the outward values of democratic societies from office desks with calculators and statistics instead of guns. Gretton shows that much like Nazi officials, these individuals do their job in a system that allows them to blank out the costs of their decisions through a myopic focus on tasks and growth charts. The system promotes the fragmentation of the mind and protects its beneficiaries. To a greater or lesser extent, we are all part of this "banality of evil" (Arendt, 2006) that by now affects all life on Earth.

Most people are concerned about climate change and want their children to have a safe future. Through a process that Weintrobe (2013) calls *disavowal*, cognitive knowledge is kept separate from felt experience, with the result that there is little experience of urgency.

Compartmentalization maintains an emotional equilibrium. The seeming "normality" of everyday life has a barely hidden malignancy to it. Desensitization and fragmentation shield the psyche from overwhelm but inhibit mobilization in the face of unprecedented danger.

In order to stop the left-brain-dominant culture from the destruction of the basis of life, McGilchrist (2010b) calls for a greater emphasis on right-brain attributes in all aspects of culture and society, for only the right hemisphere knows that both sides are needed (p. 71).

The endeavor to reintroduce relational and systemic values into every part of society would mean an engagement with the paradoxical aspects of life that do not fit into linear thinking, the ability to acknowledge uncertainty, chaotic networks of relationships, living systems thinking, embodied practice, and other strategies that are suspiciously viewed as unscientific. It would mean a willingness to travel into the wilderness of everything that we have "othered" and allowing it to unravel the reductive story we have told about the human condition and the world. We may need to pay attention to the exiled parts in our inner and outer worlds and make an effort to notice the unseen—that which has been part of our story all along but that we haven't been trained to notice.

I believe that the mental health professions have an important part to play in supporting this shift. In order to do so, it is important to investigate where the mental health professions perpetuate a fragmented left-hemisphere-dominant culture themselves and where they collude with a paradigm that upholds an exploitative relationship with the living world.

I therefore recommend that we critically examine where psychological theories and assumptions collude with aspects of the neoliberal paradigm that is costing us the Earth, and we consciously shift toward an interconnected, relational stance within the profession. I have written about this elsewhere (Bednarek, 2018).

Climate Change against the Background of Collective, Intergenerational Trauma

I suggest here that pockets of Western culture are built on layers of unprocessed collective trauma and that the resulting fragmentation within society has had detrimental effects on its ability to respond to the climate emergency.

Little attention has been paid to the cultural and personal impoverishment that may ensue from the loss of reciprocal connection with the living world. Glendenning (1994) calls this rift between humans and the environment "original trauma" (p. 57) and suggests that accumulated suffering over generations contributed to the experience of feeling exiled from one's own humanity. She describes how isolation from the world resulted in numbing and a pervasive sense of not belonging, which has been completely normalized in Western society. She says, "I maintain that a traumatized state is not merely the domain of the Vietnam veteran or the survivor of childhood abuse: it is the underlying condition of the domesticated psyche" (p. xiii).

We only have to go back eight to ten generations to find relatives in the lineages of contemporary white Europeans who experienced the breakage in the connection to Nature, to the cycles of the seasons and traditions that provided communal containers for collective experiences. The inheritance of the last century has added unspeakable atrocities to an already traumatized field with the mass killings of the First World War; the inhumanity of the Holocaust; the terror and barbarism of the Second World War; the use of nuclear bombs on people and places; the brutality committed in the name of empire; the Cold War; the violation and raping of millions of women; the degradation of human dignity through slavery and institutional racism; the continued suffering caused by genocides, ecocides, wars, and natural disasters; the oppression of people and ecosystems in the name of progress and profit; violence committed on the basis of gender, class, race, or an idea of supremacy over the natural world. How do we attend to the effects of this much collective suffering?

And the degradation of life is ongoing in a world that faces mass extinctions while keeping the machinery of capitalism going at full speed. Many people work long hours in meaningless or underpaid jobs, producing and consuming goods and services that nobody really needs. Surely humans are not meant to live for weekends and occasional holidays, raise their hands quietly to be allowed to speak, sit indoors on a summer's day, and drown out the beauty of the world with neon lights. There is a deadening in this much Soullessness. Weller (2015) speaks of a chronically anaesthetized society where the split from Soulful living has become normalized. Could this "malignant normality" (Bednarek, 2019) be the symptom of a wounded world? If we consider the possibility of a culture or society having a collective psyche, as Jung (Jung & Jaffe, 1995) suggested, many symptoms are then not an individual's dysfunction but a sign of a suffering culture.

In general, a trauma response is an adaptation to overwhelming experiences without recourse to sufficient support (Rothchild, 2000; Van der Kolk, 2014). Without adequate external and internal support structures, traumatic experiences cannot be assimilated and are split off in an attempt to numb the wounded part and protect survival (Fischer, 2017). Protection from traumatic overwhelm can appear in the form of deflection, denial, rationalization, fragmentation, dissociation, or numbing (Forner, 2017).

Epigenetic research shows that external and environmental factors are passed down through the generations through genomes. Transgenerational epigenetic inheritance of stress, trauma, and fear has been shown to reverberate across multiple generations, even in animals, and even without the presence of the original stimulus (Berger et al., 2009; Short et al., 2016).

If societal trauma stays unresolved, the following generations are born into a trauma field that can rarely be named. It is simply experienced as "normal." Without the cultural means to digest collective trauma, any new experience that falls on already traumatized ground cannot be integrated. The threat of climate change and all the interrelated crises that intensify through feedback loops are therefore likely to sit on top of all other categories of unresolved personal, cultural, and intergenerational trauma in the collective field. In this "traumasphere" (Woodbury, 2019), our innate abilities to respond to obvious dangers are inhibited.

The trauma theory of structural dissociation (Van der Hart et al., 2006) describes how the brain's structure of specialized hemispheres facilitates disconnection in times of overwhelm, allowing the left brain to take over control and perform the tasks of daily life. The left-brain hemisphere is able to stay focused, remain positive, and function in a logical, task-oriented way, while much of the function of the right hemisphere is suppressed, remaining in survival mode, braced for danger or frozen with fear (Fischer, 2017). The functions of the fragmented parts are often disowned and experienced as "not-me." This compartmentalization allows people to be informed about climate change while staying emotionally undisturbed and unresponsive. The more reality is systematically fragmented in this way, the more anxiety builds up unconsciously, and the need for further distortion increases.

In a traumatized culture, only a certain aspect of society is free to develop, whereas parts of the culture remain frozen, fragmented, or hyperactivated and reactive (Harvard Medical School, 2019). The fight-flight-freeze response is a neurological reaction that leads to hormonal and physiological changes in the face of threat. Posttraumatic stress occurs when overactivation of the nervous system becomes chronic (Levine, 1997). Life is lived on high alert in constant survival mode. The flight response is linked to anxiety, panic, and hypervigilance;

the freeze response shows in numbing and dissociation; and the fight response can lead to increased aggression, violence, and reactivity. On a collective level, different pockets of society will have different coping strategies, depending on their resilience and support levels, but a chronically traumatized community is more likely to be reactive rather than responsive.

Defense mechanisms provide protection from overwhelm and are therefore relatively stable and immune to change. Traumatized individuals or collectives are often unable to change their responses according to rational will, even if this lack of response leads to greater suffering (Van der Kolk, 2014; Levine, 1997). Chronic trauma cannot be resolved with logic, pressure, demands, advice, or pleas. It is therefore highly unlikely that climate information and statistics can mobilize the frozen or irrational parts of society. Only if we apply a trauma lens can we see that what looks like a lack of care may in fact be an unconscious adjustment to chronic trauma.

Trauma is a relational and contextual response. If the defensive compartmentalization collapses without adequate support to assimilate this loss of protection, there is a risk of emotional dysregulation and hyperactivation, in which the higher functions of consciousness are no longer fully available. Anger, aggression, and illogical reactivity can erupt in an attempt to restore the status quo. Alternatively, dissociation and numbing or anxiety and panic are likely to increase.

I suggest that these responses can be expected in individuals and different pockets of society if defense mechanisms collapse without sufficient psychological resilience and support.

Trauma therapy suggests validating the adjustments that once were necessary for psychological survival in the absence of adequate support, however dysfunctional they may have become over time (Van der Kolk, 2014; Fischer, 2017). It is not cognition, but relationship, community, and a reconnection with safe belonging that brings healing. I therefore suggest that collective forms of trauma-informed support are needed in order to increase psychological resilience in the population. It is beyond the scope of this article to discuss how this may look in practice, but I will make some general suggestions below.

There is an emotional range within which most people can contain intense emotions without either dissociating and numbing at one end of the spectrum or going into blind panic and reactivity at the other. Trauma therapy aims to widen this "window of tolerance" (Siegel, 1999), supporting the ability to reintegrate fragmented parts and bear difficult emotions while staying connected and grounded.

Collective trauma inevitably manifests in many people's individual lives, but collective patterns of wounding and dissociation are hard to identify without theories that allow a wider perspective. In line with the dominant individualistic paradigm, most psychological theories look through the lens of a personalized psychology, often attributing mental health symptoms to a dysfunction in an individual's own personality or parental upbringing (Bednarek, 2018). Only if we extend the lens beyond isolated events or a focus on parental shortcomings (often decontextualized from their own psychosocial situation) can we see a larger picture of intergenerational and collective traumatization—a web we have been born into. I suggest that a psychosocial lens is needed to account for the extent of collective suffering expressed through individual lives.

It is beyond the capacity of most people to process collective trauma on their own. Collective trauma needs a collective container for collective healing. But most communal traditions have been lost or neglected. There is no doubt that individual support is beneficial, and at times a larger perspective is needed to mend collectively what has been torn apart. Maybe we can put a larger container alongside individual support and take part in rebuilding communal support structures, which acknowledge interconnectivity and move beyond the story of a separate self.

John E. Mack, a professor of psychiatry at Harvard, calls for an expanded psychology of relationship that includes the natural world. Mack does not believe that a mere threat to survival will be enough to create this new relationship without a fundamental revolution in Western consciousness. He suggests that psychology "reinfuse [itself] with the imprecise notions of spirituality and philosophy, from which it has so vigorously and proudly struggled to free itself in an effort to be granted scientific status" (Mack, 1995, p. 284). Values that transcend individualism are needed to stop industrialized nations from milking the world dry, asking instead what the current situation is calling for and finding the strength and the resilience to rise to it.

The research of German ethnopsychologist Juergen Kremer (1998) revolves around what he calls "recovery of Indigenous mind," which for him does not relate to adopting knowledge and customs from other cultures, but instead describes a painful process of remembering the collective stories of one's own lineage in order to go forward. He suggests that the path to a sustainable future needs to move through the historical wounds suffered and perpetrated by oneself as well as one's ancestors and what they passed down through the generations. He stresses that we cannot leap out of our historical situation. We have to bear the discomfort of turning toward the unbearable and work our way through it in order to be free from its long shadows.

At times fear and suffering accumulated over generations can be overwhelming, and sections of society may freeze into collective shock and trauma. If collective support can be brought into this contraction, many can move through despair into a wider perspective and reintegrate what has been split within themselves, their families, and the collective culture, between the generations, and between humans and the living Earth.

Once the fragmented, exiled, and marginalized parts of an individual or a society are genuinely listened to, stories of heartbreak are likely to reveal themselves. At that point, shame often emerges. Shame holds us to account for the rupture that our actions or nonactions have caused (Erskine, 1994). The ability to stay with the discomfort is a necessary stage in the process of traumatic growth. Relationship, embodiment, community, and belonging soothe the nervous system and promote reconnection with what has been "othered" and marginalized.

I suggest that the integration of personally and culturally disowned parts is a necessary step toward climate action. After all, a culture that discovers what is alien to itself simultaneously discovers something about itself and its resources (McGrane, 1989). I propose that in this endeavor it is paramount to include the more-than-human world (Abram, 1997) in the marginalized and suppressed aspects of society that need to be heard. Can we include rivers, forests, mountains, salmon, and viruses in our idea of community? Rarely is the loss of attachment to Nature discussed as a traumatic event, even though this is a loss that runs so deep that it has changed who we believe we are (Bednarek, 2018).

Integration work requires attunement to diverse perspectives, safe containers for social dialogues, and the processing of collective experiences. Literal and metaphorical re-membering of what has been dis-membered is necessary in order to bridge the divides that have been created and to face the times ahead with the resources, strengths, and vitality that are not available in a fragmented state.

Necessary Suffering

In the following I explore the idea of a certain degree of pain being the price to pay for the passage from reactivity into mature response-ability, meaning the ability to respond maturely.

Nothing stays forever. At some point, our species can expect extinction. But surely there is a more dignified way to die than sleepwalking toward demise while taking other cultures and ecosystems down with us.

With the dominant left-brain lens, concepts of suicide often seek linear reasons to explain the death drive in our species so that treatment, prevention, and control can be worked out. The explanatory concepts include masochism, self-destructive tendencies, confused thinking, internalized aggression, and death wishes. If we apply a less linear reading of the fantasies and drives of a suicidal individual or culture, we open up the imaginal space of Soul (Hillman, 1996) that demands that something or someone has to die metaphorically, not literally.

In many myths, a person who embarks on a journey into the unknown sustains a loss or a wound that changes them dramatically. In their suffering, they discover that there is a much deeper truth that underpins the appearances of their everyday life. The hero or heroine matures and then returns home with a gift in service of their community (Campbell, 1949; Rohr, 2012). These myths point out that a maturational process entails "necessary suffering," caused by the loss of the self or identity that was so carefully constructed in the first half of life. What has to die is a sense of grandiosity and the omnipotent illusion of control. Certainty has to die in order to make way for a much wider view: "In the first half of life, the negative, the mysterious, the scary and the problematic are always exported elsewhere. Doing so gives . . . a quick and firm ego structure that works for a while. But such splitting is not an objective statement of truth! Eventually, this overcompensation in one direction must be resolved and balanced" (Rohr, 2012, p. 148).

It was Jung (Jung & Jaffe, 1995, p. 340) who first talked about the two halves of life. While he acknowledged the importance of a healthy ego structure as a strong container, he emphasized that the preoccupations of the ego need to eventually give way to the much deeper purpose of Soul. The ego needs to learn to be in service to the Soul's calling, which cannot be satisfied with individualism and linearity. Soul speaks the language of longing, dreams, and imagery, and it transcends the dualistic divide between "me" and "not-me" (Hillman, 1996).

From that perspective, what looks like dying can largely be experienced as a falling into a vaster and deeper life, where the Soul has found connection to a greater whole, a relationship with the limitless immensity of life. A certain level of pain and the ability to bear it can therefore be the vehicle that allows the crossing into maturity and the growing into abilities we never thought were possible.

The attempt to cheat life by excluding death, suffering, and the intimacy of grief in the name of saving people from discomfort rarely ends well in mythological traditions. According to Jung (Jung & Jaffe, 1995), much unnecessary suffering is caused because people will not accept the necessary suffering that comes from being human. Necessary suffering allows the heart to break open to the world. On the other side of heartbreak lies a matured and ripened Soul, not in spite of the pain but because of it.

In alchemy, this is the blackening nigredo stage (Jung, 1980), a process of transformation, without which we cannot get to the essence of life. This alchemical stage of disintegration can unfold only if contained in a strong vessel. In other words, the breaking down of structure needs psychological containment. Paradoxically, a maturational crisis requires individuals to master two opposing things: to provide a strong holding container so there is no collapse into premature death or psychological breakdown, and at the same time to allow rigid structures and values to crumble and dissolve.

A time of crisis comes with the responsibility to reimagine habitual ways of doing things. If we dare to live with loss as an adviser instead of an enemy, heartbreak will travel with us each step of the way, but we also have the chance to develop the maturity to meet life with integrity, including the coming of death.

Four Lectures for the Kyiv Gestalt University in a Time of War

Peter Philippson

Before the invasion of Ukraine started, I was invited to present four webinars for the Kyiv Gestalt University. With the changed situation, we had to decide how to make them appropriate for the present context.

I sat with this for some time. Of course what I said had to link to the war, the anger, and the fear, as well as the mutual self-support and resilience Ukrainian people have shown, which has impressed people all over the world. At the same time, I felt that it would be wrong of me, sitting in my safe place, to lecture Ukrainians on how to be in a war situation. In terms of supporting the traumatized, injured, and terrified, psychotherapists and psychologists in Ukraine probably know much more than I do. So how do I do this in a way that feels honest to who I am and who you are?

I am clear about what I bring. As well as skill, knowledge, and experience in Gestalt theory and practice, I bring my background as the child of refugees (from Germany, Austria, and Poland in the 1930s), much of whose family died during mass killings, both as Jews and as antifascist resistance fighters. I heard my mother's stories as well as her and my father's struggles to create a life for themselves as refugees in a new country, with a new language and culture, having to make a whole new set of friends and finding new work possibilities, while grieving what and who they had left behind. They (and I) also had to face hostility toward a German-speaking family living in a Britain that had been so badly hurt by Germans. I saw the struggles of my wife's parents to accept that she had fallen in love with me, and to learn how to relate to me and my mother. I am indebted to them for the care and love they still managed to show me.

So I will speak of what I know that might have some parallels with what you are facing now, if you are in Ukraine fighting or surviving, or have left Ukraine. And I hope that some of what I have to say will be useful.

The Dangerous Nature of Human Beings: Aggression and Creativity

Human beings are about the most aggressive species in the world. Other species kill, mostly to eat, or in battles over land and power or sexual dominance, or for play. Human aggression includes all of that, but adds something else: creativity. We, seemingly uniquely, have the capacity to imagine a world, an artwork, a way to get somewhere, and our aggression has the capacity to turn that imagination into a physical reality. We blow up mountains to get materials to build homes and offices, or if they get in the way of where we want to go. We kill or displace people who do not fit into our vision. There is nothing personal about it—our attention is on the vision, not the destruction caused by the vision. So the danger in our aggression is not our anger toward other people, but our visions and plans that ride over others.

Resilience, Self-Support, and Trauma

Resilience is "the process and outcome of successfully adapting to difficult or challenging life experiences, especially through mental, emotional, and behavioral flexibility and adjustment to external and internal demands" (APA, n.d.). People who are resilient do not experience less distress or anxiety than other people. They have developed healthy coping mechanisms that allow them to remain calm in the face of challenging times, often even emerging stronger than they were before. Gestalt therapy has been speaking since its beginning about this positive ability as something to be achieved by therapy, rather than therapy being about fixing what is broken; we call it "self-support." Perls commented that without there being an "other," self would have no meaning, and it is the mutual emergence of self and other in contact, on the contact boundary, that he makes central (Perls, 1978). Self-support inevitably involves not just the human organism, but its environment.

Self-support (or resilience) has two complementary aspects. First, there is the early assimilation from the kind and engaged presence of parents, neighbors, teachers, and others, with the complementary effects of both the presence and absence of this. Not only will the child be assimilating a sense of the world as pleasant and homelike, or painful and rejecting, but their brain will be developing neural connections that support that shape. After the first two years of life, that kind of extreme early shaping only changes with difficulty. This is what traditionally gets referred to as the "internal" aspect of self-support.

Then there is the support that people develop in their present-day world, which can include family, relationships, friendships, hobbies, living space, and the other-than-human environment. This is what traditionally gets referred to as the "external" aspect of self-support. But these two are not separate by any means.

People who tend to see the world as mainly friendly and rich in possibilities will usually actualize their environment in a way that mirrors this, so that when things go wrong they have many resources to draw upon. What is more, their instinct will be the inbuilt human one to join together to aggress on the problem and to support the person. Meanwhile, people whose early experiences have led them to see the world as dangerous and rejecting will usually actualize their environment to mirror that, so that when things go wrong they do not have people and resources to go to. Also, they are less likely to use those resources, even when they are available, because their understanding is that they will be either let down or betrayed and that the painful situation they find themselves in is just how the world is, and they cannot expect anything else.

The third aspect of self-support, which is often not talked about in therapeutic discussions, is that the people in the second group, who see the world as dangerous and rejecting, will tend not to be people who others want to become close to. This is because their hunger is to receive, and they usually have nothing to give. In addition, they tend to be rejecting and are sometimes dangerous in their own right. They are better at resentment than kindness and are always on the lookout for betrayal and abandonment, which actually leads to the abandonment they are expecting. I think this is happening on a national level in Russia.

These considerations will play a part in the response to the current situation.

The attachment and neuroscience researcher Allan Schore (2003) asked this question: what makes posttraumatic symptoms more or less severe following a trauma? At first, he checked the correlation between the severity of the trauma and the severity of the symptoms, but he found no correlation. The best correlation he found was with the person's attachment history. People with a poor attachment history suffered posttraumatic stress, whatever the severity of the trauma. In contrast, people with a good attachment history did not suffer posttraumatic stress; rather, they found support for themselves from both "inside" and "outside."

I wonder what Ukraine's secret is, being so resilient in the face of both its history and its present experiences in a way that has inspired people all over the world.

The basic definition of trauma is that it is the response to an event that is outside our normal range of what we have resources to cope with. It is not about something painful or frightening, if those experiences are within our range. With trauma, what happens is that this new, painful, and unfamiliar event cannot be assimilated into the person's life and becomes a separate system. We are built to do this by what is called state-dependent memory, learning, and behavior. Traumatic flashbacks are sudden shifts into the traumatic state-dependent system, including its sense of confusion and helplessness. So the work is to build a bridge between the traumatic system and the rest of the person's functioning so that they can bring their resources to the experience rather than flipping into a separated state.

This is why "trauma debriefing" is harmful rather than useful. If the person "goes into" their trauma experiences, they go into the traumatic state, where by definition they cannot access the support to assimilate it. If you are going to build a bridge, the first support pillar you build is on the most solid ground, not in a swamp. Then you build out from there, making sure you do not put too much strain on that first pillar.

To be in the room with traumatic horror stories, having our mirror neurons connected to a terrified client, can be profoundly destabilizing for the therapist, who needs to develop their own support and "anchor points," including a satisfying life, good supervision, and a comfortable working environment. Otherwise, the therapist gets pulled into the swamp. That helps nobody.

Having developed that, there is a need to go slowly toward the trauma, while still having the priority to stay firmly attached to the here-and-now therapeutic contact. It is fine to stop the client if either they are losing contact (talking into space, stopping breathing, closing their eyes) or the therapist is losing contact (avoiding hearing or looking, stopping breathing).

I remember one time when a woman client who had been a child prostitute was telling me about extreme abuse she had undergone. At times I had to ask her to stop telling me because I could not hear more at that moment, and she needed me present. She told me later how important it was that she did not have to look after me.

I often have a cup of tea that I am drinking as part of holding onto normality while going into the traumatic material. And it is vital to keep in mind that it is not the telling in itself that is important but the assimilation.

Of course, it is difficult working with trauma in a war situation where the trauma is shared by the therapist. In this situation, it is particularly important for the therapist to attend to their own grounding and pacing. Last year I did online supervision with a group of trauma counselors in Beirut, with riots and fires happening outside their homes. Almost the only thing I did was find what resources the counselors could access. One woman had completely lost hope, but I saw on Zoom what a beautifully furnished room she was in, so I asked her to move her computer around so we could all appreciate her taste and surroundings. She could access her pride in having created this, and it made her more available to support and contact.

The development of self-support or resilience is about the building of a pillar of connection, and then building out from that to the more swampy side of the traumatic experiences, making sure that neither the therapist nor the client ever loses the connection to the pillar.

Hate

I want to speak about my experience of hatred from my family situation. I grew up hating Germany and Germans for what had happened to my family during the Nazi regime. I would not go to Germany, and if I saw a film about the Nazi period, including *Cabaret,* I felt very disturbed by a wish to kill somebody. At the same time, I was a child in a German-speaking family in Britain, shortly after Britain had been in a vicious war with Germany. The hatred that many people in Britain felt for Germans sometimes led to violence and discrimination against me and my family, even though I felt that same hatred myself. It was difficult for my future father-in-law, who had been paralyzed as a soldier in the Second World War, to accept that his daughter was getting married to someone from a German-speaking family. It was wonderful that he was able to get past that and make a connection with me and my mother. Later on, I did go to Germany and found a country that was now very different from the one my parents had faced and that offered to reverse the Nazi removal of German citizenship for anyone in the Philippson family who wanted it. I am now a German citizen as well as a British one! I did not see that coming.

Some years ago, I was sent the first issue of a new journal: the *Journal of Family Violence*. It said that it aimed to explore the "problem of why we hurt those we love." I looked at this with some curiosity, and I commented to a colleague, "Who else would we hurt, the postman?!" The point was that most hatred comes in areas of intimate connection, even as a shadow side of love.

No wonder Freud wrote about the Oedipal conflict, and Melanie Klein (2017) wrote about the splitting between the "good breast" and the "bad breast," and the difficulty for a young child to hold the loved, satisfying mother and the hated, unsatisfying mother in one person. And that then moves into a wider field as the child gets older, whether it is between different groups of children in school or teenage gangs and the like.

In most family situations, the hate is balanced by the love and connection between people, so the hurting is not common and not serious. Hate is often not acknowledged, as people play "happy families." But when families break up, hatred and competitiveness can come out fully.

I saw that many years ago in former Yugoslavia, where a number of different countries and cultures had been put together as a single country. When that country came apart, the positive national cultures had become less important, and the defining point of nationhood was: "I am not like you."

Let us see what Perls et al. (1994) say about hate: "To these settled passions the other functions of the self are sacrificed; they are self-destructive. To hate a thing involves binding energy to what is by definition painful or frustrating, and usually with diminished contact with the changing actual situations. One clings to the hateful and holds it close" (p. 125).

Many times I have seen clients who have not been able to individuate from their childhood families because they remain tied to their parents by bonds of hate that survive even the death of the hated parent. This is the truth behind the otherwise meaningless platitude to forgive parents. Just to forgive leaves the unexpressed anger retroflected into self-hatred and internal angry words. But to stay with the hatred is very self-destructive and prevents the person from growing up. To forgive parents feels like a betrayal of the hurt child.

HOW DOES THIS APPLY TO WAR AND CONFLICT?

Hate has its uses. My hatred of Germany and Germans performed many functions: it supported me in fighting against fascist groups that were becoming active again in Britain; in a strange way, it aligned me with feeling British; and it was a distant and therefore safe object of my anger with people who were nasty to me because of my Germanness. Hate dehumanizes the other and makes it easier to kill, injure, or torture them. It is the easiest way to motivate conscripted soldiers as opposed to professional soldiers, whose primary motivation is an allegiance to their military unit and comrades (and who would tend to respect other soldiers even if they are fighting them). It is also a quick and easy way to motivate a population. And, of course, in the Russia-Ukraine war, Russia is providing a lot of incentive to hate by the extremity of their actions.

However, hate—while an effective tactic—is a very dangerous strategy. There is a Chinese proverb: "If you fight dragons, you become a dragon." As the Perls et al. quote cited above reminds us, hate binds you to the other and makes it more difficult to see the whole situation. It does not allow for any exit strategy, because that would be a surrender to the hated other. We can see that with Putin, who would rather destroy Russia, and maybe even use nuclear weapons, than appear defeated. But that can apply to Ukraine as well. You cannot truly negotiate with the hated other.

This is not me telling people in Ukraine to agree to negotiate away part of their territory. But at some stage, either there will be an agreed strategy for moving away from continuous war, or there will be extreme actions like nuclear attack to quickly end the war. A continuing "eye for an eye and a tooth for a tooth," as the Bible says, eventually leaves everybody blind and toothless.

Let me take an example that shows both sides of this. The situation between Israel and Egypt has been one of continuing hatred, leading to war on several occasions and acts that have stoked the hatred on both sides, including the original removal of Palestinians from their homes to refugee camps. A breakthrough seemed to come from the Camp David negotiations between Sadat and Begin, facilitated by Carl Rogers. Rogers wrote an article with Richard Ryback about the process where Begin and Sadat slowly began to discover each other as human beings rather than hated objects, and an agreement emerged from those negotiations. However, those agreements, and that move away from hate, separated the leaders from the people they were there to negotiate on behalf of, who were still hating. Begin was voted out of office and Sadat was assassinated. The success that Rogers and Ryback wrote about was a success in the wrong context. And yet, something like that has to happen, or we will continue to have war.

The more hatred is used as a tactic, the more likely it is that even army initiatives to end the war will be opposed by the population as treason and betrayal. Hate is so energized that it becomes the dominant figure, putting collective mutual support into the background. It is a figure without an exit strategy.

Hatred is profoundly destructive of community resilience. It leads to an erosion of trust and careful watching for traitors and spies. Of course, there are people in the community who identify with the enemy, but they rarely cause great trouble. Their main value to the enemy is that they lead to paranoia and internal conflict. The doctrine of "The enemy of my enemy is my friend, and the friend of my enemy is my enemy" leads to very dangerous alliances (as the Americans found after they armed and trained Osama bin Laden against the Russians) and civil wars.

In speaking about hate, I am aware I am treading into hot territory, and that it would raise many emotions in people. I thought that it would not have been honorable for me to ignore a topic that seems so important in the present situation for people both inside Ukraine and those who have left.

Life and Death

I want to start by describing an experience. I am standing in my aikido class, facing another student who is holding a sword. When he is ready, he will attack me. I feel very alive in many ways. I know that if I do not move, I will be injured or killed by the attack; the attacker trusts me to defend myself and is not holding back. My life is in my hands, and the present moment has an edge as sharp as the blade of the sword. I am focused. I do not know of any other situation where I go so quickly into a meditative state. My focus is on the moment. I do not try to guess what the other will do in the future, because if he does not do that, I am unprepared for the actuality. In Perls's (1992) terms, I am response-able in the present moment. I am aware that my life is one pole, and that the other pole of death is always present as a possibility. Death is part of the ground of the figure of life.

Our present self only remains stable as we stabilize it. To stay the same in a changing field is in itself always an act of creativity. When people come out of depression, they usually have images and fantasies of dying. And that is what is happening: the depressed organization—not just internally, but of their whole field—has to die, and a new self emerges with a new organization. With this viewpoint, the statement "I would rather die than stay like I have been" becomes a powerful support for change rather than a statement of suicide.

Conversely, the death of an organization by which we are what we are is a destruction of the self. When our world changes as it has for every Ukrainian (and with at least ripples for everyone in the world), the way we recognize ourselves is lost. People have to rediscover themselves as fighters, refugees, prisoners, and people close to death. And we do not know until we self-actualize in this new organization what kind of soldier, refugee, or prisoner we will become, which will also depend on the responses of those around us. Who would have thought that my mother, a German-speaking secretary, would end up as an English teacher and then as a painter?

I saw very clearly during the COVID-19 pandemic that the people who did very badly were those who either tried to hold onto their self-concept as it was before the pandemic or who spent their time resenting the situation. Those who did best were those who reasonably quickly embraced the new situation with an openness to how to be. The question for those people was how to balance risk and opportunity, how to balance the risk of dying with the risk of not being alive. What were they willing to give up, and what would they not give up even if it could kill them?

I want to suggest that, on this view of self, time, and space, nobody is dead; and at the same time, we are all constantly dying for who we were in the previous present moment! I am thinking of an epileptic friend who told me that epilepsy was something that happened to other people, not to him. The body shaking on the ground was not in his experience. And I believe this fits perfectly with Gestalt self-theory. Because if self is a comparison with other, for somebody in an epileptic fit, and for a dead body, there is no otherness, and therefore no self. And the organization that was that person's life does not just disappear. The ripples of our actions and intentions flow on after we die and are part of the ground of what comes next. Of course, what has changed is that we can no longer change our minds and add new actions.

There is an experience of moving toward death, as there is an experience of moving toward epileptic fit or sleep, or coma. But death is by definition not part of that experience. Death, as we relate to it, is a narrative. It carries the projection of my wish to stay alive, or my wish to die and get out of pain, or of my beliefs about the afterlife, or my mourning of the people I am separating from, or my sense of their mourning for me. It is a living organization in the time before death that opens up new vistas with each turn, fighting to stay alive, accepting that I will die, thinking of events in my life, stray thoughts about "I'll never know how my children will grow up." And sometimes all this happens, but the narrative goes somewhere different, and the death does not happen, or does not happen how I expected it to. Or my narrative is about going on living and I get hit by a car.

I have had some of these experiences. I had cancer around forty years ago, and the treatment failed. I knew that the alternative treatment had a very poor survival rate, and I went to the hospital and refused it. I was clear that I was going to die as a result. However, they told me that they were going to offer me an experimental new treatment, which they thought would work better, and nobody on the new treatment had died so far. Well, it worked, and I had to come to terms with not dying the death I had previously come to terms with, but also with aspects of my body not working as they used to because of the treatment. And the fact of expecting to die and then surviving changed me. And my friend with epilepsy died during a fit but would have had no more sense of it than of the fit itself.

Yet, just as it is the awareness of the other that gives us our sense of self, it is the awareness of not being that points up the amazing fact of our being alive now. After I realized that the cancer would not kill me, every day has seemed like an extra gift.

One of the things that become very figural in times of war is the inherent randomness of our lives. Which building will be hit by a missile, who will be shot at, who will step on a mine? Of course, we had to face that in a different way before that with the pandemic: who will become ill, who will recover, who will die? How much do we risk ourselves and those around us? This is true for all our lives, and we all know people who have suddenly had heart attacks, car crashes, sudden illnesses. When we turn left or right, we know our lives will be different, but we only know what actually happens down the path we take.

Randomness is an important part of freedom. Freedom is at the intersection between the unpredictable and the predictable. If our lives are completely predictable, there would be no choices and no freedom, and we would be clockwork machines. I would say that intelligence and consciousness would be very unlikely to evolve in such a universe because it would have no function. If our lives are completely unpredictable, again there would be no meaning to making choices, and consciousness would be unlikely to evolve. Consciousness has meaning only on the edge where unpredictability and predictability meet, that is, in a chaos-theory world where order emerges from chaos and chaos remains the background and the engine powering the order. All the orders we make in the world, whether they are about nations or self-concepts or narratives, can quickly become transparent and show the chaos underneath. But—and this is vital—that does not mean they are meaningless; in fact, exactly the reverse. What is fully fixed and ordered has no meaning, and our actions in such a world are machine acts. Our existence is the provider of meaning.

It has not been easy for me to find what to say, and sometimes it has not been easy for you to hear. I hope what I say will be something you can use in these strange and terrible times. Thank you, Inna, for the opportunity, and George for the translation. If you have any ongoing responses out of your assimilation of what I said, I would be happy to hear them. My best wishes to all of you.

When the Familiar Collapses

Living and Practicing as a Psychotherapist during a Time of War

Inna Didkovska

Translated by Valentyna Pshoniak

On February 24, 2022, Russia attacked Ukraine, and war broke out in my country. Suddenly, I was the same Gestalt therapist and Gestalt Institute director, but now in a war situation. As a lecturer, I began teaching Gestalt therapists and counseling clients who were experiencing war themselves.

The war touched me in an extremely close way: it literally broke into my house. Today, the whole world knows about the Ukrainian cities of Bucha and Irpin and about the "Bucha massacre." Both my son and I were born and brought up in Bucha. My parents lived in Bucha all their lives. This is where their graves are. For ten years, I attended school every day in neighboring Irpin. I know each street and many of the people who live there. I knew firsthand what was happening in my hometown from neighbors and friends, long before it appeared in the news.

I am sharing this so that you can get a sense of the personal and global context in which I am writing this.

All Ukrainians, including therapists and clients, are currently exposed to trauma. As therapists, we face the urgency of working with war trauma, while being affected by it ourselves.

Thanks to the support of the international community, I was able to implement two projects this summer:

- ✦ Inside and Out of Trauma: a project working with war trauma

- ✦ Existential Gestalt: a project, launched with Peter Philippson, that supports Ukrainian Gestalt psychotherapists who are experiencing war and who are working with clients experiencing war at the same time

In this article, I will share my thoughts on topics that have become figural for me in this difficult time, both professionally and personally. These
are themes of meaning and meaninglessness, responsibility and choice,
justice, agency, the collapse of background and support, and the search
for resources.

Meaning and Joy as a Means of Support

For me, one of the most difficult experiences is the experience of meaninglessness. The war aggravated and deepened this personal struggle,
which is why I want to start here.

I recently read *The Choice* by existential therapist and Auschwitz survivor Edith Eva Eger (2018). At times it was hard for me to imagine how
Eger managed to make sense of the suffering that she and her family went
through. Later in life, she met Viktor Frankl, studied existential therapy,
and worked as a psychotherapist.

Eger's first two clients were Vietnam War veterans. Both clients had
paralyzed limbs and a prognosis that they would never walk again. The
first client lay in bed in a fetal position and cursed everything in the
world: the government, God, his life. The second client was wheelchair
bound and met her looking forward to being taken outdoors. He kept
saying: "How wonderful! Now I will be able to see the faces of my children. I will smell. I will be able to share important moments of life with
my loved ones!"

Eger suggests that we all consist of these two veterans, and she highlights the importance of not getting caught in either polarity. This means
allowing oneself to feel anger, despair, and grief while also paying attention to gratitude and a focus on what is there rather than what is lost.

Following the Gestalt idea of polarities, I move from one polarity to
another, from meaninglessness to meaningfulness; from experiencing
injustice to experiencing gratitude; from sadness to joy. Of course, I would
prefer not to experience a polarity that includes despair and depression.
Even if I tell myself "You need to notice what is here and not just what
is absent. Notice the things you *do* have," this war teaches me that it is
not always possible for me to get out of bed. I experience moments of
deep despair, strong anger, even hatred and complete impotence. As of
this writing, however, after half a year of war, moving along the spectrum
between polarities, my sadness, hatred, and despair haven't transformed

into joy; but I experience relief and gratitude for what I have, despite all the loss. And I hope that over time the experience of joy will return too.

As I said earlier, I have a hard time experiencing a sense of meaninglessness. It seems that in moments of joy, people do not tend to question the meaning of life. I do not mean to suggest that the meaning of life is to experience joy, pleasure, and happiness, but joy certainly is an important form of support. When you experience joy, meaninglessness recedes.

The pain and sadness that Ukrainians face is so great that for many people the opportunity to experience joy is inaccessible. There is a sense that while people are dying in their hundreds every day, there is no time and place for joy. People postpone pleasure and joy for a period after the war has ended. This is how they cut off their vitality, lose access to the very possibility of experiencing joy, and as a result, find themselves frozen in the polarity of despair, depression, and traumatic experiences.

In my opinion, it is crucial to be able to move between the polarities of despair, disappointment, and meaninglessness on one side of the spectrum and joy, meaningfulness, and gratitude on the other.

The polarity of joy is important not only for civilians, but also for those who are in active combat. Many people who are on the front lines say it is important for them to fight, knowing that in the rear people keep living a full life, filled with sadness and meaninglessness as well as pleasure and joy. The knowledge that they are protecting a world that has joy in it can therefore be a support.

I remembered how my mother forbade me to express joy when my father got cancer, even though my dad never wanted me to lose my joy. As a matter of fact, my capacity to experience joy gave meaning to the last days of his life and served as his support.

Just as fear for a loved one can be more intense than fear for oneself, joy for others can be stronger, too, and can serve as a support and an antidote to a sense of meaninglessness. Of course, I do not suggest being invested in either side of the polarity, but it is important to maintain continuous access to the full spectrum of experiences whenever possible.

Existential Questions as Support

Following the ideas of the existentialists, life is meaningless, unfair, imperfect, finite, and lonely, especially at the point of choice. Existential philosophy became widespread on the eve of the Bolshevik Revolution and the First World War, which put an end to the hopes for a golden age. We Ukrainians had the illusion that Russia would not invade our country. For me, the situation in Ukraine and its effect on the wider world stage resembles the beginning of the twentieth century. This contributes to my experience that we are lonely, though responsible for our choices; we are finite and live in an unfair and chaotic world. Reality, even a chaotic or meaningless reality, is always more supportive than the illusion of order, justice, infinity, determination, certainty, or the possibility of avoiding loneliness. Reality can be relied upon, even if it is unpleasant, unhelpful, and not ideal.

Perls talked about how we idealize the world and then suffer when reality hits home. When I rely on reality (in my case, the reality of war), I regain the function of choice and can assume responsibility for my life. I can respond to what is actually going on. Some people choose to defend the country with weapons, some volunteer or provide other assistance, and some flee the country. Those who chose to stay continue to give their children a Ukrainian education, the opportunity to learn the Ukrainian language and to live in their country without getting caught up in delusions. Whatever my choice, I am less likely to sit and wait passively for an escape. Accepting reality in all of its unfairness, imperfection, and finality opens up choice and the orientation toward support. Understanding that the world can be unfair gives me a foothold. It helps me assume responsibility for my life, for justice, and for what is happening around me. Although I have to admit that the position of "it's all their fault" can be very seductive at times.

When I rely on reality, I accept that a war is raging in my country and that my life is affected by this. What is important is what I chose to do next.

Russia attacked Ukraine. Russia is responsible for its actions. But the victim ceases to be a victim when they accept responsibility for their "response-ability" and understand that they need to defend and protect themselves as well as ask for help.

When a woman is raped at gunpoint by a soldier, she is the victim. At the same time, becoming a victim of an event or a person does not mean remaining a victim and living like a victim forever. It is the reliance on reality that makes it possible to step out of that role. I have worked with victims of violence for ten years, and I recognize the difficulty in taking responsibility. It is tempting to take the stance of: "I am a victim and everyone owes me." It is important to legally recognize that violence and damage have been done and that there is a need to atone for the damage. On the other hand, it is crucial not to develop self-destructive behavior on the level of individuals, groups, or the country. Ukrainians are likely to struggle with this victim syndrome for some time.

When facing reality and relying on it, it becomes possible to take responsibility for your life; be less dependent on circumstances, or recognize your dependence on circumstances and, as a consequence, suffer less from them; and to take care of your safety, because no one else will take care of it. This does not mean it is mandatory to leave the war zone or the country. After all, soldiers on the front line also care about their safety. It is important for this choice to be conscious and based on reality. As Gestalt psychotherapists, our goal is not to ensure that people will save their lives or flee from war. It is important to respect a person's choice and responsibility. We can support contact with reality rather than supporting fantasies about how life should be. I can choose to give my life for the freedom of my country, for my freedom, or for the freedom of my loved ones. Therapists need to make sure that this choice is conscious, that a person is not stuck in a victim role and that they create their life in response to the situation they are in.

War Trauma and Existential Issues

In the first months of the war, clients and psychotherapists faced existential issues, and many Ukrainian institutions, including ours, organized acute trauma and crisis support. This is why Peter and I created the Existential Gestalt course for Ukrainian psychotherapists.

War trauma, like all large-scale injuries, inevitably raises the entire list of existential issues I mentioned above: lack of meaning (meaninglessness), loneliness (choice), finiteness (death), responsibility (freedom of choice), lack of justice, imperfection, and so on. These issues affect individuals, groups, and the nation during war.

War trauma is collective in nature and can be described in terms of:

+ a loss of agency at the individual level and at the level of social groups (people do not have any influence on war, death, bombing, or large-scale violence)

+ a loss of control

+ a loss of a sense of security (even in a safe place)
+ an attack on the self (identity)

In my experience of living and practicing as a psychotherapist in a time of war, numerous issues become figural: the collapse of the background, the collapse of support, the collapse of personality, and the loss of connection with meaning. All these phenomena can be viewed as systemically interrelated, with each one affecting all the others.

Background Collapse

The collapse of the background in the case of war trauma happens because it is not just one connection that collapses, but the entire familiar system of belonging at the local, communal, and national levels, such as social networks, professional contexts, and local embeddedness.

When a loved one dies, you go outdoors and see that everyone keeps living while you grieve. You may think: "Why is everyone doing well when I am not?" At the same time, the psyche is supported by things in the background remaining stable. The peculiarity of war trauma is that the collapse of the background occurs for everyone throughout the entire country, for every neighbor and friend. The entire support system falters at the same time. The world seems to be collapsing. This is scary and makes it harder to reach out because everyone needs support, and others may have an even more difficult situation than you.

The Collapse of Support

The collapse of supportive pillars is another phenomenon of war trauma. When working with trauma, the search for resources and support is essential in order to avoid further traumatization.

What support is there? I distinguish here between external and internal support systems, while also recognizing that this division is clumsy because internal support emerges, exists, and changes as a result of contact with the environment.

External support includes close and reliable relationships as well as the entire environment. In particular, I think of:

+ personal, family, professional, and seasonal *routines;* for instance, in Ukraine it is important for everyone to sow the fields and plant vegetable gardens in the spring and harvest them in the fall

+ the *professional environment,* with its relationships, contacts, and routines

+ *the opportunity to stay at home,* which is illustrated by the proverb "your home is your castle"

+ *the opportunity to stay in your native country and its customs;* in Ukraine, we say "it is easier to walk on one's own land"

+ the *natural world* and *other living beings* can be a source of both support and further grief (e.g., needing to leave pets behind or seeing agricultural land destroyed)

As I write, the war has been going on for six months and is not likely to end soon. We have the difficult task to build support while our background is collapsing. People in towns and villages around Kyiv that survived the occupation in the spring are rebuilding houses without the certainty that they will not be destroyed again.

Before the war, I often worked with the consequences of trauma. Now we have to work with the impact of trauma that continues to happen every day. People do their best to restore some supportive pillars and resume the continuity of life even in the conditions of an unending war.

At first, the war was experienced as an acute situation of high intensity. Over time it turned into a chronic situation of high intensity. We are adapting, just as we adapt to extreme heat or cold. Initially, it seems that you will not survive in such heat, but after three months you adapt, and even a slight coolness makes you shiver. The same takes place in trauma. At first, you do not know how to get through it. Then you build new support, you adapt, and soon it becomes more difficult to deal with a calm environment. The price of such adaptation is desensitization. This is a creative adjustment.

Through trauma work, it becomes possible to regain sensation. Pain usually returns first. I compare this to coming inside from extreme cold. You immerse your frozen hands in warm water, and along with the blood flowing through the vessels, the sensation of pain emerges. Despite the pain, the return of sensation is important, as it reduces the depth and duration of PTSD.

As therapists, we need to work carefully with the psyche's ability to desensitize. Although Gestaltists work with the assumption that in ordinary life sensation is a support, in a situation of chronic and acute trauma, it is desensitization that can become a support. Depending on the adaptive abilities of each individual, we must strike a balance between decreasing and increasing sensation in our work.

Self-Support as Self-Reliance

Self-reliance or self-support is an ambiguous topic. Gestalt therapy tends to avoid splitting a person into "I" and "myself." On the other hand, self-support is also environmental; it is directly linked to the experience of support by others, which affects one's ability to create a supportive environment and make use of available resources.

For example, when I was growing up, it was very common in Ukrainian culture for support and concern to be expressed in the form of criticism. Now, when something goes wrong, instead of looking at myself with love, I criticize myself. An internal monologue or dialogue takes place. Depending on one's internalized experience of support, this can either sound like "You are a loser, you will fail" or "Be patient, my love."

This internal dialogue is a product of past experience. The environment developed this form of self-support or self-sabotage. The ability of restoring the collapsed background is therefore directly linked to one's early attachment experience.

The depth and duration of PTSD depend on the level of support available after the trauma. It also depends on how a person's basic attachment was formed. If a secure attachment has been formed, the person is more likely to have access to self-support, even when the background of the habitual support system has collapsed. With insecure attachment, a person is more likely to see rejection everywhere.

For example, many Ukrainians are returning to Ukraine, despite the fact that the war continues and it is still not safe. People are returning even to insecure cities in pursuit of familiar footholds as they have struggled to settle in a new place. The challenge of recreating support systems in a new place can be too overwhelming. That is why many people are returning and rebuilding their houses amid an ongoing war.

Personality Transformation

Collapse and transformation of personality is another significant phenomenon of war trauma. I notice it in myself, as well as in my work with clients and students. Most of my Ukrainian clients say that in peacetime they would never have made the important changes they made in the context of war. Both clients and therapists share that their self-image has collapsed, or continues to collapse. The flexibility of self-functions, in particular personality, and the ability to creatively adjust to a changing environment often ensures the resilience of a person in difficult, existential situations. My ideas about myself and my meaning-making are therefore changing.

For example, before the war, I used to invest energy in online and offline educational projects about relationships: male–female relationships, family relationships, relationships with children. Now, this has almost lost its meaning for me. This does not mean that meaning will not return, but everything that happens outside the context of the war loses its meaning for me. And even the topic of family relations is now devoted to family relations in the context of war.

Another example of the transformation of my personality is related to the fact that before the war I considered myself apolitical. I felt it was important that psychotherapists take a nonpolitical stance and be as neutral as possible. This has changed. Now I have a definite political position. I will talk about it, as I have in this chapter. It determines the clients I choose to work with and which students I take on as trainees. The range of my acceptability and tolerance has significantly decreased.

I view these changes in the light of self-deconstruction and self-transformation, in particular with regard to personality and ego functions. I have been experiencing this self-deconstruction and self-transformation for several months and will probably continue to experience it for a long time, like many other Ukrainians.

With these reflections, I would like to conclude my commentary. I would like to thank Peter Philippson for his lectures on existential Gestalt and Steffi Bednarek for inviting me to contribute an article to this book. As director of the Kyiv Gestalt Institute and a psychotherapist who works and lives in a time of war, my background and familiar support mechanisms have been partially destroyed. This opportunity to write has turned out to be a very big and new supporting pillar for me. I sincerely hope that this article will provide support for the Ukrainian psychotherapeutic community, just as Peter Philippson's course Existential Gestalt became a supportive pillar, according to the feedback of the participating therapists.

Chapter Six

Personal Journey: From the Consulting Room into the World

Moving Out of the Clinic Space

Intertwining Psyche, Community, and World

Sally Gillespie

From childhood, my love for the living world and my desire to help people were central strands in my life that ran along parallel tracks, balancing and supporting one another but always separate. It felt natural to make psychotherapy my career, to relax in the garden and bush, and to vote Green. After days of sitting inside, interacting with the living world was how I restored myself. Being a Green voter affirmed my values of caring for people and ecosystems. But then in 2006, these separated strands started to intertwine when I became the president of the C. G. Jung Society in Sydney. I had just read George Monbiot's *Heat* (2006), which opened my eyes to the immensity of systemic change needed to effectively address global warming. My immediate response was to write about and then chair a panel on the psychological dimensions of climate change for the Jung Society. This then became the springboard for the Jung Society's publication of *Depth Psychology, Disorder and Climate Change*, edited by Jonathan Marshall (2009).

I may not have done much more if it were not for a vivid dream experience I had one night in 2008. In this dream, I swung on a rope suspended high above the Earth. Looking down, I saw the whole planet with its land masses heaving and continents being engulfed by rising seas while millions of people in the oceans desperately attempted to cling to fast-disappearing shorelines. I knew in my gut that I must join them. Letting go of the rope, I dropped, becoming one of many trying to cling to heaving shores.

I woke deeply shocked. Any possibility of distancing myself from climate reports collapsed as I shook for the vulnerability of all beings on Earth, and the realities of collective fate. While I did not believe my dream was prophetic in a literal sense, I did feel that this was the end of my world as I had known it. My dream crashed through denials and rationalizations, rupturing foundational beliefs about personal autonomy and independence. I knew that, in one way or another, I would spend the rest of my life engaged with the realities of looming climate breakdown.

As my mind and life became immersed in climate science and politics, I was grabbed by the question of what was happening psychologically for me, and others like me, as a result of being engaged with this existential threat. Did this ongoing engagement inevitably lead to chronic despair and overwhelm, or was it possible there might be a journey of psychological development within this experience that fostered resilience and consciousness change? If so, what did this process look like, and what support was needed to facilitate it? Grappling with the ebb and flow of my own fluctuating emotions, denials, disavowals, and apocalyptic imaginings, I sensed there was an unexplored world to be mapped by tracking the psychological experience of engaging with climate issues on an ongoing basis.

I started this mapping through journaling and dream working, while yearning for companionship, stories, and shared reflections to explore what was essentially a collective experience. I found that the world of climate activism generally had little space for emotional reflection or expression. What I was looking for was something more akin to the dream groups I had facilitated or participated in over many years, but with a focus on the shared experience of climate engagement, rather than on personal issues of healing and self-development. This led me to envision a participatory research group that would commit to observing, sharing, and analyzing the feelings, thoughts, dreams, and images that arose in us in response to climate issues. When I was writing my PhD proposal in 2009, there was no such term as "climate psychology." As I spoke to various universities, explaining that I wanted to use my psychotherapy background to research the psychological experience of being engaged with climate issues, I was met with both interest and puzzlement. Who could supervise such a thesis? No one felt qualified. Fortunately, the social ecology department at Western Sydney University, which valued diverse, community-based research, took me on, providing me with a harbor from which to sail into uncharted waters.

While I was passionate about doing research, I also felt committed to my therapy clients, most of whom suffered from the effects of complex PTSD. As a Jungian psychotherapist, I often highlighted the healing presence of the natural world in my clients' lives, dreams, and sandplays. The threats of the climate crisis, however, stayed largely in the background of their lives and minds as they grappled with the complexities of negotiating traumatic effects and daily life. For some time, I juggled psychotherapy work and climate psychology research and writing, but as time went on my body and dreams communicated the toll this was taking. Deep down, I knew I was burnt out as a therapist, but my ego was not willing to surrender its identifications or its sense of responsibilities and achievements built up over many years. In the end, recurrent bouts of bronchitis and the needs of aging parents became my personal tipping point. I was forced to recognize that I was unable to be the reliable and consistent therapist that my clients needed. Urged on by increasingly dramatic dreams of falling to Earth, I wound up my private practice.

While illness pushed my decision, I was well aware it was not just my body that was having difficulty maintaining a steady presence in the therapy room. After decades of attending to the inner lives of myself and my clients, I questioned how healthy an individual life could be when embedded in a world of ecological and climate breakdown. James Hillman famously observed that after a hundred years of psychotherapy and analysis, "people are getting more and more sensitive, and the world is getting worse and worse" (Hillman & Ventura, 1993, p. 3). I too perceived that downplaying ecological and social realities left individuals and communities impoverished and ultimately weakened. I also questioned what had been the effect on me personally and professionally of being so immersed in individual narratives of trauma. Had this unbalanced focus contributed to my own burnout? My growing immersion in an ecosystemic worldview was dismantling my belief in Western-style individualized psychotherapy as the best practice for a society to address issues of mental illness. Instead, I wanted to be a part of developing a more culturally aware, systemic, and grounded basis for safeguarding mental well-being, however that might look in a climate-disrupted world.

Nevertheless, thirty years of studying and practicing Jungian psychology was not something I could simply abandon. Nor did I want to. Through the process of planning and conducting participatory research, I began to sort out what I wanted to bring out of the therapy room and into the larger community. Fittingly, one month before the first meeting of the research group, I dreamed of clearing out my old clinic space. While doing so, I told a new therapist working in this space about a dream I'd had, in which I had been told that "when you find the cure for cancer you've always been looking for, then you can come back." Within this dream, a therapist in the same office then told me about a dream she'd had in which she was told she can work with a dream in two ways: external and symbolic. To do this, she had to strike out her first lot of associations and then start again.

I found this dream, and the two dreams within it, affirming and intriguing. First, there was the recognition that my lifelong drive to alleviate suffering and illness needed to be taken out into the world to be worked with in a different way. Viewed through a symbolic lens, cancer is a systemic disease requiring systemic treatment; it wreaks havoc through unchecked growth, undermining the capacity for life. The symbolic resonances with climate change were clear. At a more literal level, I wondered if there could ever be a cure for cancer, and therefore my need to work outside the therapy room. On many levels, it seemed, this dream pointed to the need to develop ways of working in the world that grappled with existential fears and unbalanced systemic conditions of life and death, both collectively and individually. Second, the guidance that dreams can be worked with by paying attention to both the literal and the symbolic, and by implication internal and external life, validated my invitation to research participants to share dreams in group discussions. In sharing dreams that felt connected with the research topic, participants could comment on whatever aspects of the dream resonated for them, personal or collective.

I knew from years of facilitating dream groups that dreams could identify and illuminate sociopolitical dynamics as well as personal and interpersonal dynamics, as other dream researchers have observed (Bulkeley, 2008; Lawrence, 2007). While my dream groups generally backgrounded sociopolitical insights in favor of an individualistic focus on healing, my own experience told me that collective and individual perspectives needed to be consciously held in relation to one another rather than being positioned as a binary, both in dreams and in life.

I felt supported in this approach by depth psychologists Mary Watkins and Helene Shulman (2008), whose work on the psychologies of liberation points to vital directions for all psychologies to consider in this era of systemic crises and injustices. Their call for psychologists and others in the helping and research professions to cultivate critical and reconstructive approaches rests upon the relinquishment of individualistic perspectives and an expert outsider stance, in order to support the creation of communities that consciously nurture the well-being of all of their members and ecosystems. Embracing this collaborative, systemic approach means holding individual and collective understandings and resources side by side, while learning to sit with others through the uncertainties and discomforts of change.

My participatory research group's reflective conversations and dream-sharing in response to climate change awareness proved to be productive and rich. Values, ways of knowing, mortality awareness, worldviews, political polarization, and consumer culture were all grist for the mill, along with a gamut of climate emotions. This fueled a group process that accepted and normalized climate emotions while challenging binaries and conditioned worldviews, as I have written about elsewhere (Gillespie, 2019). Like other research participants, I recognized how the group process facilitated a maturing process in me, strengthening my capacity and commitment to climate engagement. In theorizing about this process, I identified the development of an intertwined psychological, sociopolitical, and ecological consciousness as a crucial element for sustaining climate engagement by supporting emotional intelligence, critical analysis, and collective action (Gillespie, 2022). As climate change became climate crisis, I understood that this development of systemic consciousness is vital for contemporary culture as a whole to support the repair of communities and ecologies ravaged by extractivist practices and neoliberal ideologies (Weintrobe, 2021).

After I finished my doctorate in 2014, I looked for ways of working that were accessible and that supported the development of ecopsychosocial perspectives. One direction had already revealed itself when I presented a paper on my research at Psychology for a Safe Climate's (PSC) inaugural conference. PSC had been founded in Melbourne in 2011 at the same time as I was running my research group, but this conference was my first meeting with them. I was beyond excited to at last find colleagues in the field of what was now being called climate psychology. While many of the papers presented at the conference had an academic focus, mainly directed toward explaining climate denial, what emerged from the audience was the need for psychological support for climate activists and policy makers. When I presented my research data along with a symbolic data map of an imaginary world that named the topics and themes of my research group's conversations (Gillespie, 2019), it resonated strongly with participants, who immediately wanted to work in small groups to map their own climate emotions, thoughts, and experiences. Fortunately, there was space to respond to this request. We spread out sheets of paper and art materials for small-group work so participants could talk while drawing a map of an imaginary world with geographical features named through sharing their thoughts and feelings (e.g., Cape Confusion, Mount Militant, Sadness Springs).

After this conference, PSC pivoted its work toward supporting climate activists, scientists, researchers, policy makers, and teachers through reflective conversations, art therapy processes, and mindfulness practices. Working as a facilitator with them and with other organizations, I have facilitated many small-group reflective conversations, some combined with the use of art materials, sandplay figures, and natural objects, as well as with dream-sharing. There is no lack of need or opportunity for this kind of work, which has now become widespread through the Climate Café movement and other online groups. One of the challenges with this work for the facilitator, however, is generating income. Being older and economically privileged, I have been able to largely work pro bono, but I am very aware of the conflict younger colleagues have between wanting to do this work and needing to be paid for it. My hope is that as this work becomes more recognized and valued, it will be better financially rewarded.

As well as doing hands-on work, I have taken up speaking and writing opportunities. There is a thirst and a need in the general public for acknowledgment of climate distress, and for perspectives and strategies that can sustain engagement. Hand in hand with this discussion, there also needs to be a critique of the damages of Western cultural assumptions, extractivist values, and neoliberal economies that underlie climate and ecological destruction. Rather than hold my political allegiances and views apart from my psychological work, as I once did, I have learned to speak up in support of the profound cultural changes that are needed to heal the prevalent numbness, disconnections, addictions, denials, and injustices of contemporary societies.

Nurturing cultural change is the work of many voices and actions combined, the more diverse the better. In Australia, we have the privilege of living in a country that has been cared for over millennia by Indigenous cultures of immense traditional ecological knowledge and social cohesiveness. Learning from and giving leadership to all Indigenous cultures is vital, as across the globe we grapple with the devastating effects of ecological destruction and social injustices wrought by colonizing cultures. Bringing evolving psychological perspectives into this work of cultural change helps people consciously recast personal and social identity in relation to ecological realities, political injustices, and existential concerns, particularly transience and death anxiety, relatedness to Nature, living in community, and the search for meaning and fulfillment. Supportive psychological insights acknowledge and normalize the conflicts and complex feelings aroused by the acknowledgment of climate and ecological crises, contributing to public dialogues about the necessities, complexities, and uncertainties of change. Not only can psychology professionals introduce emotional intelligence and compassion into community discussions; we can also use professional privilege to invite in voices that are often marginalized. For instance, when I was asked to chair a panel for the "Six Months on from COP26" conference in 2022, I invited youth activists and community leaders from climate disaster–affected areas to speak, rather than academics or psychology professionals.

Of all the ways that climate engagement has been a catalyst for interweaving my passions and applying a systemic lens to my life and work, the most pleasurable has been to integrate gardening and outdoor time with psychological experience and political outlook. One choice I have made is to show up fully in my local community garden and permaculture food forest. Once these places were my refuge from work life, where I kept my head down and my conversations limited to soil health and compost making. But over recent years I have become more comfortable with bringing my psychological expertise into community gardening where needed. This might be by helping to ease conflicts that can arise in committee meetings or working bees, having supportive conversations with fellow volunteers who are in some way struggling, or being a public voice for the mental health and ecological benefits of gardening in podcasts, blogs, and the media. I also happily talk politics with those who are up for it, which is most gardeners these days as we contend with the increasing severity of droughts, heat waves, and floods.

Probably my biggest challenge is how to answer when people ask what I do. There is no neat pigeonhole for my work, nor clear boundaries between my work in community and personal life. It is freeing to step out of categories. When I do talk about climate psychology in response, it sparks more interest than puzzlement these days, opening the way for a connecting conversation.

Although I have contributed to the work of training psychological professionals to be climate-aware in their practices, I have no desire to return to the clinic space. Working to counter the contemporary systemic disease that undermines the health of our living world, like a cancer, is my lifetime work now. Although I am nearing retirement age, there will be no retiring for me, not only because there is so much to do but also because attending to psyche, community, and world together is an enlivening and rewarding way to live.

Moving Out of the Consulting Room and Living with the Climate Crisis

Rosemary Randall

In 2005 an old friend from the early days of the environmental movement organized a conference to celebrate his sixtieth birthday, and I presented the key ideas from a paper I had recently written (Randall, 2005). I had not expected it to be of much interest. I was writing for other psychotherapists, wondering what our profession was going to contribute to the rapidly developing crisis of the climate and offering some theoretical explorations. What grabbed the attention of the environmentalists at the conference was my description of the psychoanalytic concepts of denial and disavowal. The idea that the movement's difficulties stemmed not from a lack of good information but from the infinite capacity of the human mind to defend itself against unwelcome news was a revelation to many of those present.

I came away from that conference with two things. One was the idea that the real audience for my paper was not my fellow psychotherapists but the environmental movement and the people they were trying to talk to. The other was practical. In my bag was the CD of a program my friend had developed that measured people's carbon footprints. Its innocent questions about diet, home, travel, and income had shocked me into the realization that my average ten-ton yearly footprint was a long way from the sustainable 1.5 tons it needed to be.

I spent the train journey home thinking about what to do. My mind moved to the practical skills that psychotherapists have: our capacity to listen, empathize, and create the safety needed to explore painful issues; our attention to unconscious processes; and our ways of responding in ordinary language to resistance and defense. I was also mindful of the work I had done in groups and the way that attention to group process transforms learning and aids cooperation.

This was the start of a move out of the consulting room. With my partner, Andy Brown, I set up a community organization aimed at helping people understand how their lives affected climate change and what they could do about it. We trained volunteers in how to listen, how to hear the resistance that creeps into people's body language and voice as they watch the numbers stack up and realize that most things they take for granted—the annual flights, the car mileage, the comfy but drafty house, the meaty diet, and the comfortable income—all added up to an unsustainable future.

We took the carbon footprint program to summer fairs, meetings of community organizations, classes of adult education students, parents' groups, the Women's Institute, university societies, faith groups, and trade unions. Over a four-year period, we visited over one hundred organizations and held over three thousand one-to-one conversations with people about the impact of our ordinary UK lifestyles. The calculator produced some numbers at the end, but at the heart of each conversation was the aim of supporting someone, however briefly, in the realization that their life needed to change and that no matter what technology and policy could achieve, many of our taken-for-granteds were simply unsustainable. The relationships between technological efficiency, policy change, and personal and cultural change are complex, but there is broad agreement that people in overdeveloped countries like the UK need to make big reductions in their travel, meat and dairy consumption, and general consumption of goods and services.[1]

Most people expressed disquiet and concern. Some became upset, and some defensive. A few remained indifferent or expressed anger at us as the bearers of bad news. I recognized my own distress in these reactions, and it helped me to reflect on them. Early on we realized that about a quarter of the people we spoke to wanted something more. Primarily, they wanted support in reducing their own impact. It was in response to this that we developed the Carbon Conversations project.[2]

Carbon Conversations brought groups of six to eight people together around the common problem of our unsustainable lives. Membership of the first groups was drawn widely from our local community. Later groups were sometimes run within organizations, e.g., workplaces, among students, and in faith groups. As time went on it became apparent that many group members were relatively well-educated and that considerable work was needed to widen the diversity of group membership.

Each group member received a handbook with detailed information on what was involved in carbon reduction, placing this in the context of the cultural and psychological difficulties of change.[3] The groups themselves provided many different ways of approaching the difficult facts: explorations of the meaning of home, food, and travel; conversations about status and money; group discussions based on an appreciation of the difficulty of change; games that modeled policy options; and most importantly space to talk about what you felt. We trained volunteer facilitators in the basics of running a group and let the project loose.

To our surprise the project became national and then international as it was translated into several European languages and used in Canada, Australia, and the United States. Its appeal lay in its psychological framework and the fact that it viewed carbon reduction as enmeshed in both complex systems and complex emotions. In 2009 it won an award at the Manchester International Festival, and a booklet including a description of all twenty award winners was presented to every delegate at the Copenhagen climate negotiations. The *Guardian* described it as "one of the most promising approaches to climate change."[4]

Those were very different times. The Kyoto protocol was being implemented. The UK government was preparing and then passing its Climate Change Act. Political activists were moving effectively against the fossil-fuel industry with direct action against power stations and their supply chains. The voluntary sector was banded together in Stop Climate Chaos, making the case for an effective political settlement at the 2009 Conference of the Parties (COP) to the UN Framework Convention on Climate Change, which was expected to set the path for the following years. Community organizations at this time saw their role as helping people understand the changes that were needed in everyday life and to begin that transformation. It was a period of frightened optimism. No one I knew expected business to deliver without a struggle. No one expected government to accept the necessity of de-growth without a fight. But there was a sense that the zeitgeist was open to change.

All that came to a spectacular halt with the 2008 financial crash, the disaster of COP15 at Copenhagen in 2009, and the election of a conservative government in 2010 that imposed austerity and set about repealing environmental legislation. The 2020s are very different times. The world has continued to warm. Promises are made but not kept. Seemingly impassable thresholds have been breached. A new and deeper sense of fear infects all those who allow themselves to know what is happening. Often I find myself in the same dark places as those I am trying to help.

How should we, as psychological practitioners, respond to this? In 2021, I began a conversation with Rebecca Nestor about whether Carbon Conversations could be revived. Rebecca is a board member of the Climate Psychology Alliance and was involved in Carbon Conversations in its early days. What needs would a revived project have to meet? What, if anything, from the old project might still be useful? We were meeting young people who were desperate about the future, older people filled with guilt at the way they had ignored the crisis for so long, and activists worn out and demoralized from years of campaigning. Could we use the effectiveness of the facilitated, small-group model to respond to the needs we were now seeing?

The result is a new project, Living with the Climate Crisis,[5] hosted by the Climate Psychology Alliance and drawing on its members' skills in group facilitation. Daniela Fernandez-Catherall, a psychologist with expertise in collective narrative practice, joined our team, and together we have developed new materials aimed at the issues that now feel crucial. The project builds on elements of Carbon Conversations and on other psychologically based workshops on climate change we have run. The groups are designed to be run across a period of eight to ten weeks, with two facilitators and six to eight group members.

It is a cliché that the antidote to climate despair is action. Far less attention is paid to the process of moving from a state of acute distress, anxiety, and grief into a form of action that feels commensurate, practically possible, and sustainable over time. This is the process that the Living with the Climate Crisis groups aim to address.

The project has three core themes:

+ facing into the painful feelings that the crisis induces, and finding strength through group connection;

+ engaging with what we call the "ecosystem of change"—the multiplicity of possibilities for action from personal carbon reduction to direct action;

+ learning to communicate well about climate issues, both publicly and with those close to you.

Dorothy Stock Whitaker (1985) coined the phrase "using groups to help people" to describe psychologically facilitated groups that brought people together around a shared problem, and this describes Living with the Climate Crisis groups well. Living with the Climate Crisis groups are neither therapy, nor education, nor action, but they have elements of all these. They walk a delicate line between them, offering possibilities of healing, learning, and working together. Broadly stated, the purpose of the groups is to help people live their lives well in the precarity that the climate crisis brings. Healing, learning, and becoming able to act effectively in the world are three equally important goals. By healing, we mean the process of facing into grief and finding new meaning. By learning, we mean developing skills that will make us effective in the climate movement. By becoming able to act effectively, we mean making connections with others and building collective strength. Acknowledging pain, sharing stories, and talking about strengths and skills are some of the ways we go about achieving these.

In creating the groups we have drawn on a wide range of thinkers and practices. The first meetings draw partly on the new understandings that come from climate psychology and partly on collective narrative practice. We draw on William Worden's framework for loss and grief (Worden, 1983), making space for people to talk about how they feel, emphasizing that it is possible to reach a place beyond the worst feelings, and sharing ideas about what helps. We work to build collective strength in the group by using Ncazelo Ncube and David Denborough's Tree of Life method (Ncube, 2017; Denborough, 2008), which functions as a point of collective strength that group members can return to, both during the group and beyond it. The meetings on communication use therapeutic understandings of the difference between process and content alongside a translation into ordinary language of concepts like resistance and projection to explore what goes wrong in conversations with family and friends. They also draw on Marshall Ganz's "public narrative" method (Ganz, 2011) to help people use their own stories to craft more effective ways of speaking when they have an audience willing to listen, and they draw upon the work of Climate Outreach[6] in understanding messaging. Finally, the meetings on collective action root people in a systemic understanding of the processes of change, explore what carbon reduction means in practice, explore a range of actions and possible roles, and use methods of reflective practice to help people engage in the reality of what they want to do.

As in the old Carbon Conversations project, we have created materials to help others run similar groups. These consist of a *Facilitators' Guide* and a *Participants' Guide*. The *Facilitators' Guide* outlines the thinking behind the project, discusses the group work principles it employs, and provides detailed descriptions and flexible guidance on using the various activities. The *Participants' Guide* provides an aide-mémoire of what has been covered under each of the three themes and a section called "Living Lightly," which is a short guide to carbon reduction, designed to be used by "buddy pairs" that are set up about a third of the way through the group.

Two heavily oversubscribed pilot groups confirmed that our sense of how to respond to these needs was broadly correct. The materials are now available for use and are free to download at livingwiththeclimatecrisis .org. We are hoping that psychological practitioners and group facilitators of all kinds will find the materials useful and will adapt them for different audiences and different settings. The main criterion for running a group is that the facilitator has themselves been on the journey that they are helping others to embark upon. Facing your own grief, confronting your own impact, and involving yourself in action of some kind are basic prerequisites.

Living with the Climate Crisis groups is only one of the many possible interventions where psychological practitioners can use their skills, and it is important to encourage different approaches. Climate Cafés, climate circles, short-term support groups, walk-and-talk groups, climate listening benches, social dreaming groups, Joanna Macy's Work That Reconnects, and Chris Johnstone's Active Hope are just a few.[7] Having the flexibility to meet different expressions of need is important. What helps one person may not be right for the next.

For therapists, all these practices involve moving out of the privacy of the consulting room and into a more public sphere. They are a move away from the individualism of personal therapy, and they involve relinquishing a degree of control. This is not necessarily an easy process, but there are models in community psychology and collective narrative practice for psychological practitioners having a more open and collective role. To anyone wondering whether it is advisable to venture through the door and away from the protection of the consulting room, I would encourage you to be brave. When I set up a community organization, I became a figurehead for that organization and for the psychological methods I was advocating. I spoke at public meetings, took an overt political stand, and found myself answering questions on radio and television. Nothing dreadful happened, and I found that my clinical work was enriched as people brought me issues they would previously have concealed. You do not have to go as far as I did, but the crisis demands collectivity. It demands cooperation. It demands that we take a place in the public sphere. It demands that we acknowledge our shared distress and develop new ways of working therapeutically in the struggles to come.

In my own life, I have been hugely helped over the last four years by a group of colleagues in my hometown. Cambridge Climate Therapists[8] began meeting in 2018, and this small group has grappled actively with this question of where we belong and what we do as therapists. A useful concept that we shared early on was the sense that we were most useful in what we came to call the hinterland, sufficiently involved in climate action to understand what was involved while being sufficiently detached to be able to help others slow down, reflect, and pay attention to everything that makes life still worth living. Living with the climate crisis means not just addressing the desperation of the situation the world is in but finding space for continued meaning and for joy, creativity, and laughter. This experience of being in community with others is what has made me personally able to face my distress and continue to live with the climate crisis. I strongly believe this kind of experience needs to be the backdrop to our attempts as therapists to help others. We need to place the emphasis less on the "crisis" and more on the "living."

Notes

1 The UK government's current carbon reduction plans rely on UK citizens making reductions of about 30 percent in each of the key areas of their carbon footprints. See *Behaviour Change, Public Engagement and Net Zero*, available from https://www.theccc.org.uk/publication /behaviour-change-public-engagement-and-net-zero-imperial-college -london/. The Zero Carbon Britain reports from the Centre for Alternative Technology have similar estimates. See *Zero Carbon Britain: Rethinking the Future*, available from https://cat.org.uk/info-resources/zero-carbon -britain/research-reports/zero-carbon-rethinking-the-future/.

2 See www.carbonconversations.co.uk/.

3 This was later published as *In Time for Tomorrow?* Available from: www .carbonconversations.co.uk/p/materials.html.

4 See www.theguardian.com/environment/2009/jul/13/manchester -report-carbon-conversations.

5 See https://livingwiththeclimatecrisis.org/.

6 See https://climateoutreach.org/.

7 For information on Climate Cafés, see https://climatepsychologyalliance
   .org/events/500-climate-cafe-jan; on social dreaming, see *Social Dreaming:*
   Philosophy, Research, Theory and Practice; on climate listening benches, see
   https://cambridgeclimatetherapists.org/events/; on the Work that Reconnects, see https://workthatreconnects.org/; and on Active Hope, see
   https://www.activehope.info/.
8 See https://cambridgeclimatetherapists.org/.

Chapter Seven

Seeds of Change: Working with the Wider Field

Moving with Storms

Vanessa Andreotti, Rene Suša, Cash Ahenakew,
Sharon Stein, and Chief Ninawa Inu Huni Kui

Gathered in a circle, a group of STEM climate researchers is prompted to symbolically place the climate and Nature emergency (CNE) at the center of the room. They came together to examine how their relationship with knowledge (what they expect knowledge to do) and their affective investments affect their choices in climate research. Their task involves a series of exercises that will take them through a guided experience of different cognitive and affective states. The aim is to better understand the complex interplay between the cognitive frameworks and affective imprints they have been socialized into within modern/colonial systems and how these can facilitate or constrain the scope of their approach to research problems and potential solutions in relation to the CNE.

The session had started with an opening exercise that asked participants to take seven steps back. Through a series of questions participants had been invited to step back from: 1) their self-images, including hopes, fears, insecurities, and investments; 2) their immediate context and time to contemplate historical, systemic, and structural forces; 3) their generational cohort; 4) the universalization of the social/cultural/economic parameters of normality they had been socialized into; 5) familiar patterns of relationship building, problem posing, and problem solving; 6) the pattern of elevating humanity above the rest of Nature; and 7) the impulse to find immediate solutions, so they could expand their collective capacity not to be immobilized by uncertainty, complicity, and complexity.

Separability as Imposed Impairment

The session's focus was for the group to consider relational sciences and technologies as defined by Chief Ninawa Huni Kui, of the Huni Kui Indigenous nation in the Amazon. The session started with a recorded statement and invitation by Chief Ninawa himself. The statement emphasized that the climate catastrophe and biodiversity apocalypse are not technical problems but relational ones created by an imposed sense of separation from the land/planet, which is normalized and naturalized by Modernity/coloniality.

From the Huni Kui perspective, this imposed sense of separation, or "separability," represents a cognitive, affective, relational, and neurobiological impairment based on illusions of separation and superiority that have damaged our relationships with our own selves, each other, other species, and the land/planet we are part of, with deadly consequences for all involved. This neurobiological impairment, or neurocolonization, creates a dis-ease in our collective body, which manifests through symptoms of human greed, self-infantilization, hyperindividualism, overconsumption, arrogance, indifference, and irresponsibility. These symptoms are driving the destruction of ecosystems that are essential for human and nonhuman survival, like the Amazon rainforest, and placing humanity on a path of premature extinction.

In the recording, Chief Ninawa addresses the group by saying that while Western societies have developed advanced engineering sciences and technologies, which are often deployed for exploitation, extraction, and expropriation, relational sciences and technologies of respect, reverence, reciprocity, and responsibility have been not only neglected but actively suppressed due to the impact of colonial genocidal processes. Chief Ninawa highlights that Indigenous peoples have developed relational sciences and technologies to an advanced state.

CHIEF NINAWA:

We are now facing mass extinction in slow motion, and the colonial ways of organizing, thinking, feeling, relating, hoping, imagining, and being that have gotten us into this situation cannot get us out of it. The future depends much less on the hopeful and positive images that we want to project ahead than on our capacity to repair relations, to build relationships differently, and to coordinate effectively, right now, in the present. We will need to combine Western and Indigenous sciences and technologies if humanity is to have any future on this planet. Before we can do that, Western disciplines of science and technology will need to lose their ingrained arrogance, ethnocentrism, and universalism, and confront the harms they have caused or contributed to. Once that happens, Indigenous sciences and technologies can be integrated with Western sciences and technologies in a transsystemic way to coordinate efforts toward regeneration and the expansion and embrace of our social-ecological responsibilities.

Affective States Exercise

After hearing Chief Ninawa's recording, the group was asked to have a conversation in small groups about what they thought relational sciences and technologies were and to what extent they would have been exposed to them through their formal education. They were then invited to participate in the exercise of placing the CNE in the middle of the room and to go through different affective states representing a transition away from the affective states imprinted by Modernity/coloniality through their training in Western sciences and technologies, and toward the affective states made possible by relational sciences and technologies developed by Indigenous peoples, as Chief Ninawa mentioned in his recording. For each affective state, the group was prompted to locate the affective charge in the landscape of their bodies and to perform an embodied symbolic gesture. The six different affective states are described below. We invite readers to perform the exercise in relation to the CNE as they engage with the different affective states.

The first affective state is one of mastery, with a strong attachment to certainty and the futurity/continuity of what is perceived as modern progress, and it is therefore invested in solutions (mostly technical and universal market-based ones) to the CNE. In this affective state, people are driven by the potential to achieve something meaningful for themselves and useful for the kind of society they imagine as ideal. Personal investments in this state are also influenced by the potential for increased merit, recognition, and status in one's discipline or group in society.

The second affective state is also one of mastery and attachment to certainty, but certainty of a different kind. In this state, people are invested in the conviction that the catastrophes announced by the CNE cannot be averted, that there is not much that can be done, and that humanity will surely not survive this challenge. Personal investments in this state are related to the comfort (and also sense of righteousness) of "knowing the end" that placates fears associated with instability, uncertainty, unpredictability, and unknowability.

The third affective state is one of confusion, where certainties become precarious and unstable, also destabilizing one's views of the world and sense of oneself, but where the desire for mastery and certainty is still strong. This is a state where complexities, paradoxes, contradictions, and conflicting demands and perspectives become overwhelming and immobilizing, often evoking a sense of discomfort, irritability, frustration, and "nausea." In this state, people feel overwhelmed and immobilized by complexity, uncertainty, and unknowability; by the impossibility of simple solutions; and by the fear of hopelessness, powerlessness, and defeat in the face of the magnitude of the challenges ahead of us. Those who experience the unnerving discomfort of this state develop coping mechanisms that can manifest in different ways. They often respond from a dysregulated and fragile state of survival (fight or flight), where the impulse to impose coherence on complex systems is driven by the imperative to render the potential for immediate action viable. This drive originates from escapist mechanisms, aiming to avoid the emergence of undesirable emotional states.

The fourth affective state is one of pause and contemplation of both the CNE and one's internal cognitive, affective, and relational embodied landscape. In this state, the relationship to the CNE as an object of research partially shifts, as the desire for mastery and certainty is replaced by a yearning for deeper understanding of both the CNE and of oneself. In the state of pause, people can see the limits of different paradigms without feeling compelled to find universal or totalizing answers or solutions as a means of placating discomfort. In this state, people are no longer immobilized by the vastness of uncertainty, unpredictability, and unknowability, but they are also not quite ready to act yet because they are taking time to sit at the limits of their understandings and contemplate ways to think, feel, imagine, relate, and do differently. In this affective state, people learn to disinvest in innocence, purity, and perfectionism. They expand their capacity to sit with difficult and painful things, like our systemic complicity in social and ecological harm, without feeling overwhelmed or immobilized and without demanding quick fixes or to be rescued from discomfort.

The fifth affective state is one of epistemic curiosity, collective inquiry, and better coordination. This state engages with the complexity of the CNE, driven by a desire for the joy of collective epiphanies that may come from both the successes and the failures of testing hypotheses and carrying out social or technical experiments. In this state, researchers and disciplines do not have to prove their worth and secure their place in hierarchies of knowledge-worth; each researcher and discipline is seen as both insufficient and indispensable to the task at hand, and each is supported in sitting at the edge/limits of their professional and disciplinary knowledge in order to remain open to being interpolated by different ways of knowing. In this state, people are better prepared to self-regulate and to face and work through (rather than resent) uncertainty, complexity, and unknowability together.

The sixth affective state is one of relational entanglement with the CNE. Formal modern/colonial education has historically not trained—and cannot train—people to inhabit this affective state, which is associated with the advanced relational sciences and technologies developed, held, and practiced by Indigenous peoples that Chief Ninawa mentioned before. Those of us trained and oversocialized in modern systems can only have momentary glimpses of this state. In this affective state, as you are looking at the CNE, the CNE is also looking back at you, not as an object of inquiry but as a co-subject of inquiry. From this relational standpoint, the CNE is entangled with the same planetary metabolism that humanity is entangled with. It is simultaneously a separate entity looking at us and an entity that inhabits each one of us. Therefore, the temperatures and the waters rising around us reflect temperatures and waters rising within us. In this state, knowledge and collective epiphanies are not exclusive to the human intellect, and the agency and coordination of (the rest of) Nature is integral to the relational research process.

The next step of the exercise prompted participants to talk about the implications of the insights gained from this experience for academic research and training. They talked about the kinds of affective states that are encouraged and rewarded in different disciplines, the positive and negative implications of approaching the CNE through the range of affective states they experienced, and the potential problems of approaching the CNE exclusively through states invested in mastery and certainty. They also delved into the kind of research training that might support incoming researchers to approach the CNE through pause, curiosity, and relational entanglement, and the difficulties of justifying and implementing these approaches in modern/colonial postsecondary institutions. The conversation also touched upon the connection between the climate grief observed in students and the unpleasantness of being stuck in a state of confusion without having the means and training to move into pause and curiosity.

As an invitation to practice inhabiting the affective space of relational entanglement, participants were asked to perform two out of three short writing tasks without anthropomorphizing Nature (sitting at the limits of what is possible to imagine) and observing their own responses to the task itself: 1) a haiku representing the CNE as unfathomable "kin"; 2) a short apology note to the CNE that recognized the lack of respect and responsibility on the part of most living humans, including oneself, and a list of commitments to reverse that trend; and/or 3) a humorous poem that addressed the CNE as a teacher—a larger living entity trying to teach us to be less arrogant human beings and not shoot ourselves in the foot by destroying the ecological infrastructures that enable and sustain our existence. Below are some examples of these writing tasks.

HAIKU:

Humanity's dawn?
Our chance to learn under duress?
[to the CNE looking at us] What do YOU see?

APOLOGY NOTE:

We messed up. We don't know if we can fix it. I am deeply sorry for how immature and irresponsible we have been. I commit to not repeating mistakes already made. I commit to not turning away from the "shit." I will learn to compost. I will try my very best until the end.

POEM/PRAYER:

You are mighty and smart. We are small and have grown foolish and selfish. Humility is scarce. Change is hard and painful, and it requires discipline. Some of us are getting on with it, but we are mostly not very good at it yet. We need more people to sense your depth and power, but please be gentle with your teachings.

The session concluded with an invitation to take seven steps forward and/or aside (Andreotti et al., 2023), away from the conditioning of Modernity/coloniality and toward specific potential gifts of relational sciences and technologies:

Step forward/aside with honesty and courage to face uncomfortable truths: expand your capacity to confront reality, difficulties, and pain within and around you.

Step forward/aside with humility and vulnerability: shed your sense of self-importance, and recognize we are all both insufficient and indispensable to the task at hand.

Step forward/aside with self-reflexivity: learn to see yourself from others' perspectives, and learn to be comfortable with unflattering perspectives; remember, this cannot be about you.

Step forward/aside with self-discipline: identify and interrupt harmful patterns grounded in our socialization in systems that reward irresponsible and immature behavior.

Step forward/aside with maturity and intergenerational accountability: embrace the long-term project of becoming a good elder and ancestor for all human and nonhuman relations.

Step forward/aside with expanding discernment and attention: increase your capacity to navigate VUCA (volatility, uncertainty, complexity, and ambiguity) and to face the good, the bad, and the ugly of humanity within and around you without throwing up, throwing a tantrum, or throwing in the towel.

Step forward/aside with curiosity, adaptability, resilience, and joy in the collaborative inquiry process: learn to find joy in the process of moving together rather than seeking predetermined destinations.

Moving with Storms

This session was part of the Moving with Storms Climate and Nature Emergency Catalyst Program of the Institute for Advanced Studies at the University of British Columbia, in what is currently known as Canada (see Andreotti et al., 2023). The program adopted a multigenerational approach to the CNE, bringing together undergraduate and graduate students, researchers, artists, Indigenous activists, university staff, a cohort of emeritus/a scholars, and a leadership team to engage in a year-long collaborative inquiry on the theme of the CNE. The word "emergency" was used in the singular form to underscore that, in the same way that humans are not separate from Nature, climate and Nature are also inseparable from the living metabolism of the planet we are part of. The invitation to learn to "Move with Storms" emphasized the importance of intellectual, emotional, and relational flexibility, agility, resilience, and stamina in the face of the most complex, difficult, uncomfortable, and frustrating dimensions of the CNE.

Since the program recognized both colonialism and capitalism as central causes and drivers of the CNE, participants were supported in expanding their cognitive and emotional capacity to sit with difficult issues and to avoid deflection from distressing topics. However, rather than taking a moralizing approach to convince people to examine the CNE exclusively through critiques of colonialism and capitalism, the program approached the CNE as a collaborative educational inquiry, where different perspectives and approaches were also welcome. Throughout the program, participants were invited to consider how unsustainable economic growth, overconsumption, land occupation, cultural subjugation, labor exploitation, racial discrimination, and other forms of historical, systemic, and ongoing social and ecological violence have brought us to where we are today.

As the program unfolded, the importance of preparing participants to engage with the CNE as a super-wicked challenge became evident. The sense of separability that Chief Ninawa describes as a neurobiological impairment is normalized and naturalized in societies across the world, in large part through formal modern/colonial education systems, which also mostly train academics to universally apply linear logic, to expect seamless progress, and to see themselves as neutral and reliable experts and observers. This training leaves people unequipped, unwilling, and unprepared to address the CNE as a super-wicked challenge. Wicked challenges (Rittel & Webber, 1974) are defined in the Western literature as challenges that are hypercomplex and multilayered. They represent an assemblage of interlocked problems, where every problem is a symptom of another problem, and the solution for one problem creates problems in other layers. They also involve many unknowns, and they have longer, uncertain time scales. Super-wicked challenges possess additional characteristics, including the fact that time is running out; those who cause the problem also seek to provide a solution; the central authority needed to coordinate solutions is precarious, inefficient, or nonexistent; and responses are pushed into the future due to irrational discounting and ineffectiveness of existing paradigms and practices (Levin et al., 2012).

Approaching the CNE as a super-wicked challenge requires different problem posing, problem solving, coordination, and accountability strategies, including engaging with the CNE itself as an ongoing, educational, lifelong, and lifewide collaborative inquiry. In terms of creation and application of theory, wicked challenges require an approach where hypotheses and experiments are responsibly grounded in the most relevant analytical frameworks, but these frameworks are also considered part of (and subject to) the inquiry and engaged with in a minimalist, self-critical way. In terms of methodology, wicked challenges require approaches that foreground uncertainty and that prioritize abductive rather than inductive or deductive reasoning, because a large amount of the variables are fuzzy or unknown. In terms of analyses, wicked challenges require diffractive and diachronic reasoning, which enable us to hold multiple, moving layers of complexity in tension without the impulse to flatten these layers into a coherent, controllable, and predictable whole.

The collaborative inquiry approach itself requires a high level of self-reflexivity and psychodynamic self-assessment on the part of inquiry participants, whose internal drivers, approaches, and analytical frameworks are also part of and subject to the inquiry. This is particularly important in the case of the CNE, where intergenerational stakes are very high and emotional investments are intense, given the urgency of the matter. How different people experience the affective charge of the CNE inevitably affects the research decision-making process.

In modern/colonial institutions it is often assumed that research is a purely cognitive practice, but the CNE challenges us to pay attention to the affective and relational dispositions that also shape the knowledge we create and mobilize. In this sense, relational entanglement (which involves relational intelligence and relational accountability) is also about interrogating and expanding the ways we relate to knowledge, language, reality, time, place, and self. The super-wicked nature of the CNE defies desires for mastery, certainty, and universal answers that are reinforced in modern/colonial education.

While disinvesting in these desires can result in an initial sense of deflation or defeat, exercises designed through depth education principles (Machado de Oliveira, 2021), such as the ones described above, can support participants in expanding their emotional threshold to not be overwhelmed or immobilized by uncertainty, complexity, and complicity. This is necessary if we are to coordinate climate engagement, inquiry, and collective experiments across different communities and contexts with more humility, self-reflexivity, and a recognition of the partiality and provisionality of all analytical frameworks and approaches to problem-solving.

Collective Consolation

The Paradox of Climate Cafés

Rebecca Nestor and Gillian Ruch

Making Connections

Rebecca Nestor

In a local café close to my home I sit with eight other humans, around a table scattered with cups and plates. People are holding a gull feather, a beach pebble, a dried beech leaf, a pine cone, a forked stick, a seed head, a shell, and a sprig of rosemary. In my hand is a shell, about the size of my fingernail.

I look around the group to ensure everyone is ready to begin the opening round, make eye contact with my cofacilitator, and then give my attention to the tiny shell in my hand. Its fragility connects me with grief and loss, and a sense of the fine balance that is life on Earth. I put these emotional responses to my object into words, and I invite someone to make a connection in relation to the object they have selected.

THE CLIMATE CAFÉ MODEL

This is how we begin a Climate Café, following a model developed with the Climate Psychology Alliance (CPA). So what are the principles underpinning our approach, and what do they achieve?

Adapted from the Death Café model (Underwood, 2014), a Climate Café is a simple, hospitable, empathetic space where fears and uncertainties about our climate and ecological crisis can be safely expressed. Climate Cafés are one-off, advice-free spaces that normalize the discussion of climate feelings, foreground the collective rather than the individual response, and enable the breaking of the "socially constructed silence" (Norgaard, 2011) by providing a sturdy but explicitly temporary support structure.

They arose in different parts of the world, but our own experience of them began in the long, hot, dry summer of 2018 in Oxford, UK.

REBECCA:

I had noticed that more and more anxious conversations about the climate crisis were happening around me. People who had not previously talked about climate were saying, "The trees are dying." "This is worrying, it's not normal." "Climate change is here, isn't it?" These conversations were often awkward, short, bumping very quickly into silence. But they prompted a thought: what if there were a space specifically designed to enable more of these exchanges? Would they enable people to open up a little and think together about what climate change means for them in their communities? I personally felt full of restless anxiety and a wish to *do something*. My background as a climate activist and facilitator influenced my decision to try out a supported group process inspired by the Death Café model, which has a similar taboo-breaking purpose. Developing this work was a way to manage the fear that underlay my restlessness while also making a contribution that I thought others might welcome. Climate Cafés began in Oxford that autumn. The focus was not to debate climate science, discuss climate policy, or talk about climate actions, but rather to give people space to express their feelings. It was not that civil action was frowned upon or discouraged; quite the contrary. Rather, this space was not for that purpose. The people who attended in those days were mostly white, older, middle-class, and female. Despite this limitation, I could see a surprising depth to these facilitated conversations between strangers in small groups. After a year of holding Climate Cafés I wondered whether the Climate Psychology Alliance might be interested in hosting them, so I began to talk to Gillian.

GILLIAN:

For me the imperative to get involved was almost the reverse of Rebecca's. Rebecca spoke to me about how she felt drawn to offering a psychologically attuned space that in part responded to the shortcomings of activism: a place to *be* in relation to the crisis, and intentionally not to *do*. As someone who has not historically been an activist, for me the attraction of Climate Cafés was the space it afforded to *do* something in response to the crisis through the *being* approach that a psychologically attuned space offers. Over time the interconnectedness and complementarity of these *being* spaces and the *doing* of activism has become more prominent in our minds.

As our society has awakened to the climate crisis, Climate Cafés seem to have caught the imagination of many, and CPA now runs them online every month, with a short training workshop for would-be facilitators also in place. As a result, Climate Cafés are being offered online and in person across the UK, some European countries, the United States, and Canada. There is also a peer supervision group operating in accordance with reflective practice principles. Mindful of modeling the complementarity of *being* and *doing,* CPA strongly encourages facilitators to take part in supervision and is developing additional support systems. Mindful of the importance of the collective in facing the climate crisis, CPA requires its Climate Café facilitators to work in pairs.

SOME FEATURES OF CLIMATE CAFÉS: COMMUNITY, LOCALITY, TIME

It is very difficult to talk to each other about what is happening to our world. As humans, we have psychological defenses against some of the unbearable feelings associated with the climate and ecological crisis. We may also feel the moral injury (Weintrobe, 2020) and dissonance evoked by knowing one is part of a destructive system. Our "socially constructed silence" is fueled by guilt, shame, feelings of judgment, and a not-always-conscious awareness of the contradictions between our lifestyles and what is required to act.

So the Climate Café model seeks to offer a place where people can take the very first steps toward speaking about their responses to the climate emergency. There are no guest speakers and no talks. There is no discussion of policy or actions. There is gentle ritual within an ordinary setting. The group members work together to create a temporary community, generate their own content, and feel the connections between them. Our intention here is twofold. First, we want to offer a kind of toe in the water of breaking the climate silence. We think it may feel easier for people new to the climate crisis to attend a Climate Café if it is clear that they will not be forced into action that they are not ready for. Second, it reduces the risk of being inadvertently shamed—by lack of knowledge, or by not having taken personal climate actions—and of the group creating a polarized dynamic that will feel unsafe. This approach foregrounds the caring function of a Climate Café, which is an essential aspect of our collective task of attending to what Sally Weintrobe calls the "culture of uncare" (Weintrobe, 2021).

The "no commitment" temporal principle is central to the Climate Café model. Many do not even require advance booking. People are welcome to join a Climate Café once and never return, or they may find that they want to deepen the experience through repeat attendance. Combined with the "no action" principle, it reduces the threat people may otherwise feel. Despite the absence of commitment, for some the fleeting encounter may *whet the appetite* to explore further, and perhaps they will eventually come to a place where action is possible. The sensibilities that the climate crisis evokes require Climate Café facilitators to be alert to and respectful of the pace at which people can engage in climate-related conversations. For some it is an immediate relief to express their distress; for others there is a hesitancy as they encounter and explore unfamiliar ground. Defenses are an inevitable part of the experience. It is vital that Climate Cafés are experienced as ordinary but nourishing spaces where people do not feel *force-fed* and where the extent and limitations of their appetite are recognized and respected.

ARE CLIMATE CAFÉS A FUNCTION OF PRIVILEGE?

It may be particularly difficult for people of privilege to talk about the climate crisis. The defenses and dissonance described above can combine with white fragility to create an especially toxic version of socially constructed silence. This silence can lead to the erroneous belief that climate alarm is the preserve of a small minority. Does this mean that Climate Cafés are only needed in relatively privileged communities, where ecoanxiety can be described as a function of that privilege—the fear of losing the privilege and, as futurist Vinay Gupta puts it, "of living in the same conditions as the people who grow your coffee" (Hine, 2019)?

We do not think so. As people of privilege, we acknowledge the fear of the loss of privilege in ourselves; Rebecca's restlessness in 2018 (referred to above) carried that fear. But the wider reality is that forms of climate distress are experienced by people from all communities in a range of ways that are often stronger in people closer to the reality of the effects of climate change or who feel abandoned or silenced by those in positions of authority or influence (see, e.g., Hickman et al., 2021). All of those forms of distress need to be understood and shared because being alone with distress may lead to its becoming overwhelming. In people of privilege, this overwhelm may encourage the creation of a fortress mindset and prevent those in the global North from taking the action that is needed. Climate Cafés are designed to help people find words together for what is happening to us and to see the diversity of emotional responses, and through this to begin to understand it better at a social level (Norgaard, 2018). Climate Cafés need to be a collective resource, and we have worked to develop them in this way. Compared with the original narrow social group, we are now finding that the spaces are full of difference and the learning that comes from difference. Facilitators need to be mindful of the presence of white fragility in Climate Cafés and find ways to protect group members from its damaging effects.

This is not to say that people of color, who so often have experience of being more directly affected by the climate crisis, somehow have a responsibility to educate others. For example, some Climate Cafés may need to be protected spaces for people of color, and the CPA is experimenting with this approach.

THE POWER OF SOLACE

We have both been repeatedly moved by the sense of connection between Climate Café participants, whether face-to-face or online, that is created in a short space of time. While this affective dynamic is mostly implicit, it also frequently finds explicit expression at the end of a Climate Café when participants are asked to share one word. Many share the word "connected." When this happens, it can feel genuinely heartwarming that a sense of togetherness in the face of adversity has been generated so swiftly and meaningfully.

We have come to conceptualize this sense of connection as finding solace. A beautiful, seldom-used word, "solace" comes from the Latin *solari*, "to console," and it means to find "comfort or consolation in a time of great distress or sadness." It captures the sense of care and concern that Climate Cafés can generate: for each other in the Café, friends, family, loved ones, and the natural world.

THE PARADOX OF ACTIVISM

As we explained earlier, a founding principle of Climate Cafés is that they are "no action" spaces; the focus is on feelings as opposed to action. Often people who are not already engaged with climate activism express a sense of relief in the Climate Café, having felt up until that point that there was nowhere to share their distress. For climate activists, the experience of relief from their activism fatigue is equally welcome. People engaging from either of these positions can make their encounter with reality together. The hope then is that these more affectively engaged and balanced responses to the crisis will generate a more attuned and sustainable agenda for those already engaged as activists and will create the possibility for action by those who are new to the reality of the crisis. Paradoxically, then, despite its commitment to "no action," a Climate Café can be understood as an important action: a space that can contribute to reenergizing the established activist and to enabling those less familiar with climate activism to find a place to act.

Social Dreaming

Julian Manley

I had a dream a few nights ago:

> I was standing outside the front door of what looked like a stereotypi-
> cal haunted house on a dark night, clouds fleeting over a bright, white
> moon. Three stone steps lead to an imposing front door. The name of
> the occupant was on the door: "Ventre." It should have been frighten-
> ing, cold, and wet, but it was not. I walked in and continued ascending
> the stone staircase inside the house. I reached the first floor, a huge open
> space of a room with wooden planks for a floor, wood paneling, a huge
> stone fireplace, and some kind of gothic stone carvings over the mantel-
> piece. I was wondering what this place was, and I suddenly thought it
> must be the house belonging to Monty Python, and I was bound to meet
> at least three of the "Pythons" in a minute. A Siamese cat appeared, its
> eyes piercing through the gloom. I approached to stroke it, to feel its
> warm fur, an animal presence, at last. Suddenly it leaped at my legs and
> dug its claws into my trousers to climb up my leg.

I woke up surprised rather than shocked. That day, I learned that one
of the UK Conservative Party leadership contenders, Rishi Sunak,[1] had
made the ridiculous error of thinking that the English town of Darling-
ton was in Scotland. In my mind, I associate this absurdity with Monty
Python, and therefore with my dream. I imagine Sunak as the owner of
the cat in the dream and as somehow being responsible for its leaping at
my leg. Casually, I mention this on Facebook. Now my individual dream is
part of the social world. Am I, or is anyone else, any better off as a result?

This is the kind of associative thinking that occurs in the context of
social dreaming. As usual, my first thought is that this is somehow a bit
embarrassing, hardly a serious topic, and all but meaningless. Again, this
sensation is a frequent part of social dreaming. That is what I think,
but my feeling is different. I am moved and struck by *something,* and this
feeling—or "affect," more than a feeling, an embodied emotion—appears
to count for more than the thought. Perhaps it is something akin to
love, an unknown "Something" that brings the famous Beatles song to
mind; something undefinable yet vital. The tune resonates in my head.

Amazingly, I remember that George Harrison was a great fan of Monty Python and supported their films. What kind of links and associations are these?

"So what?" I hear some people ask, a touch of irritation in their voices. I think of the cat in the dream. I associate it with the magic of the cat in ancient Egypt. I think of the catlike Sphinx and how the puzzle of the Sphinx has been identified as a way of approaching the associative psyche of the social dreaming matrix, not the Oedipus of the individual dream but the Sphinx of the social dream. As Lawrence (2005) said, the social dreaming matrix "places participants in the realm of the Sphinx (knowledge) as opposed to that of Oedipus (the psyche of the person)" (p. 59).

Richi Suncat! Will he be our next prime minister?

As Prospero in Shakespeare's *The Tempest* knows, there is a fine line between hilarity and dismay:

> You do look, my son, in a moved sort,
> As if you were dismay'd: be cheerful, sir.
> After the dreams are over and vanished into "thin air." . . .
> Our revels now are ended. These our actors,
> As I foretold you, were all spirits and
> Are melted into air, into thin air.

Something of the dream remains. In *The Tempest,* this something is the love between Miranda and Ferdinand.

Human existence is as much bounded by the unconscious of sleep and dreams as it is defined by conscious awareness. In a sense, everything is a dream, as *The Tempest* suggests:

> We are such stuff
> As dreams are made on, and our little life
> Is rounded with a sleep.

So, what have we learned, now that our revels or dreams or social dreaming matrices are over? Or, "so what?" I am tempted to answer "nothing," along the lines of Oscar Wilde's saying that "all art is quite useless." All dreams are quite useless.

It is the very uselessness of dreams that renders them of use, but this is not usefulness in the sense of utility. The "uselessness" of dreams encourages the dreamer in social dreaming to reach a state of negative capability, that is, the state of mind that does not seek to find answers to everything;

rather, in accepting complexity and in the recognition that there may be more than one answer, or many answers, the mind relaxes into the possibility of living with uncertainty. Wilfred Bion, following on from the original idea of negative capability propounded by the poet Keats (1958), brought this to the attention of psychotherapy. Looking for answers or truths or facts that are inevitably based on past experiences can divert the thinker's attention away from uncertain materials. That is to say, what is uncertain will tend to be ignored or rejected as baggage that gets in the way of answers. This is the kind of thinking that goes on in any definition of scientific thought, including "social science" thinking. The reductionism of scientific thought processes exists to support a way of thinking that leads toward defensible and repeatable solutions to problems. For example, if an engineer is building a bridge, thoughts that are extraneous to this task could prove fatal. In social dreaming, however, all thoughts and feelings are accepted as valid. A dream thought from an individual may appear meaningless at one moment but may at some later stage acquire a meaning or a sense in association with other thoughts or dreams. This is what it means to share dreams and associations in a social dreaming matrix.

Another way of putting this is that social dreaming is comfortable with complexity and staying with difficulty instead of attempting to simplify and reduce the multiplicity of complex problems into easy-to-digest chunks. This is why social dreaming is particularly apt for supporting thinking around massive and almost hopelessly complex issues, such as climate change, and how to deal with, react to, and manage the disturbance and anxiety associated with such thought. Through the condensation (Freud, 1991) of dreams—that is, the amalgamation of multiple images and meanings into single scenes—and by the accumulation of meanings through the collage of social dreams in the matrix, complexity is celebrated. No longer is complexity a source of frustration encapsulated by Morton's hyperobject (2013). Instead, the shared dreams of a social dreaming matrix become a window that gives onto the hyperobject as a whole. The participant in a social dreaming matrix is in thrall to the complexity, in creative fascination with it. Without being able to explain everything about whatever is emerging in a matrix, the dreamers achieve a sense of bonded thought and an understanding that sometimes feels cathartic or even spiritual.

Social dreaming has its own language that is difficult to "translate" (Manley, 2020). If I return to my initial dream, I struggle to enunciate its meaning. Perhaps seeking meaning in the sense that this word might be normally understood is in itself a mistake. Just as all art and all dreams are useless, so dreams are meaningless in the rational-logical sense of meaning. However, even if dreams are useless and meaningless in one sense, they are full of usefulness and meaning in another. This is the terrain of the Sphinx, where knowledge is a riddle.

In the case of facing up to and even having a chance of dealing with climate change, the ability of perceiving the problems as a whole is enhanced by social dreaming. There is something of a Kleinian depressive position in accepting the enormity of complex contradictions inherent in understanding the effects of climate change. Outside the social dreaming matrix, how often is anger directed at certain sections of society, leaving other sections free of criticism? The sinners who fly, for example; the ignoramuses who eat meat; the careless who fail to recycle plastics. As opposed to this, there is the defense of the individual who can say that they do not fly, never eat meat, always recycle, or do not buy plastic, and therefore they are doing their bit for the planet. The tragic truth of this situation is that no matter how piously an individual follows this course, the problem lies in complex systems, not in individual actions. Individual actions will make individuals feel better but are not enough to resolve our climate issues. The individual action—I will not fly—is easy to articulate and unequivocal in its execution. Social dreaming is not concerned with the individual. As it is often said of social dreaming, the matrix is concerned with the dream, not the dreamer—in other words, the social complexity, not the individual simplicity.

The thinking behind individual simplicity and its deceptive clarity in itself leads to impossible contradictions, ranging from the simplest everyday experience of shopping for the organic or fair-trade or plastic-free or not-at-all banana in an independent corner shop, to the disappointment of buying an electric car (only accessible to the rich) which runs off fossil-fuel-powered electricity and on the backs of exploited labor in impoverished areas of the world. Social dreaming provides an opportunity to get away from individualistic, rational struggles with the world and to embrace the reality of ecological complexity. When social dreaming succeeds, it is also therapeutic. Many participants attest to this unintended consequence of social dreaming. The therapeutic effect derives from this encounter with the uncanny aspect of this complex reality in a way that can be understood, not through words but through the language of dreams. To use the Climate Psychology Alliance tagline, social dreaming provides a way of "facing difficult truths" that otherwise would often be too terrifying to contemplate.

Social dreaming is a training of the mind to think new thoughts, creatively and democratically. It taps into the social unconscious that is associative in nature (Manley, 2018). Thinking by association is also breaking the straight line of linear thinking. Associative thinking is rhizomatic thinking, where new thoughts are like spontaneous sparks or the sudden emergence of lightning from dark clouds and rain. Like lightning, new thoughts can be sudden and unexpected, but their import rumbles in the aftermath. Through this process, the participants in social dreaming come to a state of reverie in togetherness, where "becoming dream" is experienced as a new reality, a dream reality.

Notes

1 Rishi Sunak is a prominent UK politician in the Conservative party. At the time of writing he had lost a Conservative Party leadership competition to be prime minister. Circumstances have changed, and he is now (as of November 2023) prime minister of the UK. Such a development resonates with the sense of predictions that is sometimes associated with social dreaming.

Warm Data Labs

Steffi Bednarek and Bec Davison

Warm Data

Steffi Bednarek

The ecology of interrelated crises, like climate change, war, structural racism, inflation, geopolitical conflict, migration, and intergenerational trauma, to name but a few, is so complex that no one person or discipline can come close to understanding the constantly changing dimensions, especially as sense-making will depend on the lenses we find ourselves using.

With the urgent call for clear solutions, it is tempting to reduce the complexity of interdependent systems by breaking them down into smaller chunks, separating education from health, the economy, technology, ecology, philosophy, and so on. These fragmented aspects are then treated as multiple individual problems. Data is removed from its context and analyzed in isolation. Even sustainability projects often operate with this mechanistic mindset without spending time to consider how to reembed the information within the context of life.

Rarely do we widen the lens to take more of the interconnected (and often messy) whole into account. But information that is removed from its context only tells a narrow story and is therefore likely to lead to misinformed actions, where the solutions we find quickly become part of the problem. Unintended consequences often wreak an additional layer of havoc.

The territory we deal with is life, and life does not happen in categories and disciplines. These are maps we have created in order to navigate complexity. Alfred Korzybski's work reminds us that the map is not the territory. This may sound obvious, but in reality the confusion between map and territory is commonplace. Without including and valuing data that is relational and "alive," we miss important aspects of the territory of life. The issue is not that we are not trying hard enough. It is that we are reducing complexity instead of stretching our capacity to include contradictions, uncertainty, and changeability in our approach.

The big problems of our time will not be solved by data generation, statistical forecasts, and rational analysis alone. These approaches have their place, but we need a wider lens, one that tells a bigger story of how we are "lifeing" together. Being human is a transcontextual participation in a bigger whole. Each person is a complex, living synthesis of all the manifold influences that are continually shaping us. We cannot be reduced to a single story. You may be a psychotherapist, but you are also a body made up of more microorganisms than human cells; you may be a parent, a musician, a partner, someone who struggles with addiction or depression and who may have suffered great loss . . . the information we hold about this wild, moving, and changing synthesis is "data," but it cannot be measured with instruments. It is warm, human, living data that changes in contact and relationship. Just because it cannot be counted, that does not mean it does not count.

The term "Warm Data" has been coined by Nora Bateson, founder of the International Bateson Institute and daughter of the influential systems thinker Gregory Bateson. Her work continues and furthers her father's work and reminds us of the importance of information that is woven into the fabric of constantly evolving ecological systems.

WHAT IS A WARM DATA LAB?

A Warm Data Lab is a kaleidoscope of personal stories and interrelationships that aims to widen the complexity that is collectively addressed by a group.

Participants are invited to step out of familiar labels and compartmentalized roles and step into connection with each other. The Warm Data facilitator starts the lab with a deeply personal, transcontextual story that is explicitly relational instead of educational. Then, small and constantly changing groups reflect upon a given question through multiple lenses, which usually include the economy, family, community, health, religion, technology, ecology, and more. As we all already carry a rich transcontextual synthesis within us, people are invited to tap into this Warm Data pool by responding to others and the contexts from personal experience rather than abstract theory.

Through ever-increasing complexity, the focus of attention is widened beyond the usual frame of reference. These transcontextual experiences dilute the artificial boundaries we usually draw around aspects of our lives. This increases and stretches our ability to see a broader picture and to hold multiple perspectives at once, including contradictions. At the end of a Warm Data Lab, the widened perspective will not be left behind at the venue but will be carried into the way we approach life, work, our communities, and our families—not in linear and causal ways, but in a warm, fluid, and rhizomatic flow. This is a step toward weaving back together what has been torn apart.

The takeaway? *We* are the takeaway!

The Bringing to Life of a Warm Data Lab

Bec Davison

On the day I left the Warm Data training, I cried. I had a feeling that something was about to change within me, and crying was part of letting go, releasing, embracing whatever was to come. Something visceral occurred. A physical, mental, and emotional movement. A deep unlearning—hidden, embedded, rooted. Habitual thinking was expunged in what felt both liberating and painful. The moment of realization: Warm Data is not this or that, it is something other, something that cannot be named or caged.

Back at work, every conversation was influenced by the transition that had happened over those five days of training. Energy was created in the conversations, something different, something warm.

It was decided: a Warm Data Lab would be run for Preston, UK. My initial reaching out was to someone who could help with logistics (the venue, the invites, etc.) and to people with high influence: a commissioner for drugs and alcohol across the whole of Lancashire, a director of the community drugs and alcohol services, and a leader of a grassroots community group working with current and previous drug users.

Explaining the transcontextual process was vital. Applying this to the context in which we worked (drugs) gave space for the penny to drop: people are not just drug users and nothing more. Talking about community, us as community (workers), and how stuck we have become to what now seems like intractable problems.

I wrote the invite. I did not want any organizational name or logo, or job titles, included. This required explanation of my own organization and some negotiation. I used this as an opportunity to introduce them to Warm Data. It worked.

I felt very comfortable with "whatever happens, happens, whoever comes will come"—until I didn't. And then I dreamed that no one turned up. So I called Helen, my right-hand woman.

Helen provided feedback: no one knows what a Warm Data Lab is, and they are put off by "data" and "lab." Can you change the invite?

A crisis of confidence started to creep up. I felt out of my depth, naive, stupid, hopeless. I wanted to retain the integrity of the Warm Data Lab (I must call it this) while not wanting to put people off—riding two horses. A double bind?

I called Rolf, a fellow Warm Data host, who told me: "On a page, it is cold. Make it warm, Bec. Invite people so they come. The world needs this. Explain Warm Data when they're there. Then it's warm. You got this."

One of my team asked me, "When was the last time you felt like this?"

I replied, "I can't remember."

"It's good for you."

My grandpa came to mind. New things will be resisted. If you're asking people to step out of their paradigm, however dysfunctional and unhealthy it may be, there will be resistance. Roll with it.

I changed the invite. I took "Warm Data" out and provided a description of what might happen. The numbers went up.

My mind constantly wandered to the stories I was going to tell. I dreamed about stories, I thought about stories, I talked about stories. I realized I was forming principles when forming my story:

+ A parallel story to the context we are in (therefore, I couldn't really tell a work story)

+ A story focusing on community and relationships

+ A story focusing on self within the system—inner and outer, where is the edge of me/us

+ A story that tells of how I (and others) have learned to be in the world

+ A story that shows we can lean into the ecology of crises

Another dream: my grandpa turned up to guide and nudge me. *Tell a story that roots you in that place.* And so, through some iterations, I formed the story of my relationship with my grandpa, how we argued, how he taught me to argue, how he taught me that the world is complex and difficult and beautiful, and kindness and compassion will get you through.

The venue was booked: a Hindu community center with a temple. The food was ordered—hot samosas. Bunches of flowers were ordered from a local florist.

And so, a room laid out with chairs in threes and fours, multiple contexts written out in my hand, beautiful flowers all around the room, samosas, tea. I made sure every person was warmly greeted by someone as they walked into the building. And then I stood at the front of the room, turned the music off, smiled, and began telling my story:

"I was lucky enough to have four great-grandparents until I was ten years old. They all lived in Bolton, as did their parents and their parents before them, as far back as we can discover. . . ." My story was personal, it was about me, it was about how I have learned to be in the world, how my grandpa had learned to be in the world, how we had learned to be in the world together. It had no end.

And then I paused.

And then I took a samosa and used it as a metaphor to explain Warm Data. I did not talk about transcontextual processing, making the decision to only use language that as many people as possible would understand.

And then I revealed the question: "What is relevant in a changing world?"

I asked people to answer the question using the context they were sitting within. I asked them to move when they were bored or lost energy. I asked them not to get stuck in one place.

The next hour and a quarter was energized, stimulating, deep, connected, open, honest, amazing, funny, sad, hopeful, mind-blowing.

I stood again and the room quieted. I asked, "What did you notice?"

The stories, the stories, the stories.

I asked: "What questions are you left with?"

What are we going to do now?

Anything you want!

"The closest I've come to relapse in twenty years," said one person, smiling away. He felt like he had done drugs.

The power of the room setup, the greeting, the suits not wearing suits, there being no request to introduce oneself, no attention paid to role, title, status; the tonality and the expression of the story; the coming together of all of this and more cannot be underestimated or emphasized enough. There are deep reasons for that.

The diversity in the room invited a vitality. I did not actually know "who" many of the people were or what had brought them there.

As one participant said, "I have never felt so safe without being told what to do."

There were rules with no rules. Safety without saying so. Openness and honesty without boundaries being outlined. The group held the group.

Nothing was harvested, written down, recorded, collated. People connected, made plans, came together.

I see, I feel, I believe . . . the labs open up possibilities, they bring deep, human connection that leaves the existing paradigm somewhere else. It left people hungry, wanting, ready.

We need to let our edges swirl, until we realize there is no edge.

Radical Joy for Hard Times and the "Attending to Place" Practice

Harriet Sams

Radical Joy for Hard Times, or Rad Joy, is a charity whose ethos is to deeply connect people with places that have been damaged through human or extreme natural acts. The idea for the charity was born when founder Trebbe Johnson interviewed David Powless, an engineer from the Oneida Nation in upstate New York, United States. During the interview, he told her that he had the realization that the large pile of steel waste he was recycling was "an orphan from the circle of life" in that it had stepped out of the natural cycle of belonging to death, rebirth, and decay, causing pollution and toxicity. The waste had essentially been rejected from Earth systems and by human consciousness, as if orphaned. This powerful concept inspired Trebbe to found Radical Joy for Hard Times in 2009. A key aspect of Rad Joy is that, even when these places are orphaned from the circle of life, they are still very present in humans' existence: visible, tangible, telling a story of human behaviors and values, when they are usually ignored. Trebbe believed that these places need attention and honoring in some way, to return them to a living system.

When a place is hurt, the people who love that place hurt too. Turning to places that have experienced pollution or are in the process of being wounded can invoke complex feelings; often strong emotions arise, such as anger, revulsion, and rejection. Rad Joy supports curiosity and attention so that something healing can be found inside this difficult relationship, intending to lead to a love of place. Turning toward a challenging place, rather than turning away, invites personal reflections about self, community, and land as part of a relational web. Turning toward the trouble writ large on the land may lead to compassion for place and others, awareness about how we spend our money and treat what we consume, ecological responsibility, community cohesion, a sense of connectedness to a broader world of multiple species and elements, and love of land. Essentially, reciprocity and belonging weave into community a consciousness of places from which we often would turn away.

The Earth Exchange

Rad Joy initially offered one simple practice called the Earth Exchange. The Earth Exchange embodies reciprocity in one simple ritual of attendance: we go to a place that is wounded or hurt in some way. We speak to it of our feelings, our own hurt, anger, frustration, or sorrow. We express our grief, our pain. We bring whatever it is that this place is bringing up for us to share. We can also remain silent and be in quiet contemplation. Then, we begin to collect items from the place itself in order to make an *act of beauty*. This can be anything at all; it can be intuitively created and totally unplanned. It can be a collage, a mandala, a bird, anything made from what we find at the place. Then, once we have created our act of beauty, we speak of how we feel. What did we notice about the place that is alive? What other animals, insects, tracks did we see? What is thriving? What is this place, when we get to know it and spend some time in relationship to it? What do we now feel about this seemingly wounded place and the wider world, once we have turned our attention to it?

The Earth Exchange takes very little time; it can be undertaken within an hour in any location where we happen across a place of woundedness. It is a gift of reciprocity in that we are essentially saying to the place, *I see you and I am giving back something beautiful to you.* In this way these orphaned places, which are growing rapidly in number throughout the world, return to our consciousness, brought into our field of relationship.

The Earth Exchange brings four steps into this short, powerful practice. At first, we do not like what we are encountering. Then we wish for things to be different, and we experience strong emotions about this. Next, we pay attention to where we are and the land as it really is. We make the act of beauty. Finally, we notice our relationship to place anew. A shift occurs; what was our experience has changed, because we have paid attention. We stop yearning for what cannot be found in the old way of grasping for what the place is not, and we begin to see with new eyes what *is*. Within this new way of thinking lies the potential for new thoughts, processes, and acts to emerge.

There are countless examples of how this artistic practice creates a positive impact in myriad ways: activism, building resilience in participants, a significant sense of place, belonging. Yet much of this practice's potency is nonlinear and intangible. Activism *does* benefit from this practice, as it enhances the well-being of activists and protectors, offering resilience to carry on in often challenging circumstances. Communities do get connected by a sense of shared love of place, and creativity is welcomed in a sphere that perhaps would not usually encourage artistic or spiritually minded approaches to local or political activism. The land responds: places no longer become orphaned or ignored. They are welcomed back into community consciousness, which in turn means that those who have less-than-positive reasons for how they treat the land have a much harder time getting their own way. People become protectors; they have turned *toward* the wound, not away.

Attending to Place

In 2020, Radical Joy for Hard Times developed a new practice called "attending to place." This is a deeper dive into connecting to a particular place, which can take one to six months to complete. Place becomes a being, one to which we wish to make a much more significant connection. We fall into relationship with this place, naming it, learning its ways, listening to its many stories. This practice awakens an animist consciousness of place, space, and self that is nourished over months. We learn, among many things, that even a wounded, broken, and forgotten place has much wisdom to share. What wisdom are we missing by gazing at "well" lands, mighty trees, stunning waterfalls, and perfect vistas, when we can receive so much from challenged landscapes? To what are we blind and deaf? What is right here, waiting for connection with us, despite the multiple planetary woundings that we keep creating?

Why Turn Toward Challenging Landscapes?

For the last ten-plus years of my involvement with Rad Joy, I have often heard people say (and wondered myself, in times of deepest grief for the world), *What is the point*? What is the point of going to a place I would rather avoid? What is the point of sitting with a belching oil refinery, another clear-cut forest, a new housing estate, or a polluted river? Why should I make a collage out of discarded, probably dangerous and toxic stuff? What is the point when the world burns, plastic accrues, everything's dying, and the rich get richer?

For me, it's all about two things: bringing consciousness to orphaned spaces, and returning humans back to relationship with the Earth, even with the most damaged and polluted lands. Through attending to place it is possible to explore relationship with what has been taken out of the cycle of life by human and extreme ecological actions. It is also about returning oneself back to belonging once more. Through this relationship between land and human, much healing, understanding, and belonging can occur.

Turning toward a wounded landscape is a challenging thing to do. Many Nature immersion courses and ecotherapy workshops invite participants to go to pristine places and would not dream of taking people to fracking pads or polluted rivers. However, Earth is a mirror of the humanity we carry, our Earthling status both as interconnected species and as "destroyers" and "usurpers." It is wise to see what human ecology looks like by bearing witness to often immensely challenging places, as well as beautiful places. There are now skilled and experienced guides who can facilitate this by attending to challenging places. Paying attention brings us through stages of awakening, through grief for wishing it could be different, and into what is. It also brings to us a visceral immediacy of multiple ecological crises that people and animals live with; we receive deeper understanding of and feel compassion for those who cannot simply turn away from the horrors of the landscapes they have no choice but to inhabit.

Conclusion

Orphaned places are growing in number: we build, use, then discard factories; abandon shopping malls; pollute river systems; burn forests; mine and quarry for minerals; wage wars; and destroy ecosystems. No longer can we turn away from these places, because most likely they are present in our daily lives, down the street or near where we live. The urge to avoid these places is strong. Late-stage capitalist culture deliberately encourages us to go about living lives numbed to and unaware of these wounds by hiding many of the most damaging impacts from much of society's view. Rad Joy can bring a container, a process, a way of turning toward such places and gazing at the reflection it gives of ourselves. We see culture, society, and capital development mirrored upon the land; we feel the deep grief, the horror, the anger and rage because of what we see, and we offer a clear way through to the love on the other side of grief. We witness, we grieve, and we are keen for what is. Then, simply, we give back a gift of beauty. We welcome the land back into our hearts.

The land waits, just as it is. As people practice this on a more regular basis and relationships to difficult sites develop, a significant shift occurs in human-to-land relationship, a shift that is relational and qualitative, even unquantifiable and unmeasurable. Through fully turning and offering attention to these places, something within us opens. This opening brings in something unknown, even unknowable, unless you too have spent time falling into the deep wisdom of the living world, where what emerges is resolute, wise, precious, beautiful.

Tending Grief, Together

Sophy Banks

This article describes the practice of tending grief, together, and looks at why this is particularly relevant for those aware of and engaging with the climate emergency.

I learned about shared grief tending from two very different sources. Joanna Macy's Work That Reconnects includes space specifically for "honoring our pain for the world." I also encountered grief rituals led by Sobonfu Somé, of the Dagara people of Burkina Faso. Directed by her elders, who saw the degree of destruction that modern culture was wreaking across the world, Sobonfu was advised to bring teachings about ritual (particularly grief rituals) to the global North.

In both lineages there is an understanding that grief is a natural response to the conditions and experiences of life. Experiences of loss, change, injuries, disappointed hopes, and connection to the suffering of those around can lead to painful feelings, which might be called grief. People's expression of grief is influenced by the specific culture a society creates. I learned from both teachers that coming together to express, witness, and make meaning of pain contributes to the health of a community. Sobonfu might say that there is no such thing as private grief. A significant event such as a relationship ending, a conflict, or the felling of a forest affects many people. Expressing grief together allows the wider impact to be felt, understood, and metabolized.

When it comes to harm caused to communities or the more-than-human world, it often takes a group context for the (collective) pain caused by these violations to find expression and be witnessed. Shared spaces are therefore essential for the wider impacts of harmful systems to be acknowledged fully.

In the modern English culture I grew up in, the word "grief" was often used to refer only to sadness, primarily for the death or the loss of a loved one. I have learned to extend this word to include the wide range of painful feelings that naturally occur as life moves through me. In the introduction to grief work I invite a welcome for all feelings in all their flavors, including anger and fear. I invite participants to check what other emotions might be present and need tending. The list is usually long and often includes despair, guilt, and shame, as well as frustration, loneliness, emptiness, and confusion. The numbness of not feeling anything is given a place too, and so is that which cannot be named.

In relation to the climate and biodiversity crisis, I have met the diversity of feelings listed above in people's responses. What happens when these feelings are not given expression? Working in the Transition movement, I saw many groups propelled by fear and an understandable sense of urgency that was often not named, but which created a restless, driven culture. Many were on the edge of burnout. I saw groups melt down in conflict, unable to constructively process their anger at the destruction and injustices they witnessed. I saw those with more stamina enduring and perpetuating a culture where vulnerability, the need for support, and the urge to move at a slower pace were looked down on, reflecting common values in the dominant culture.

My own experience of working in response to the climate crisis gave rise to feelings of inadequacy and the sense that, however hard I worked, it would never be enough. I experienced a sense of guilt that my complicity in the system meant that my hands would never be clean. I would always be part of the systems of destruction that I was working to end.

There are many ways to avoid feeling grief, many of them highly socially acceptable within the dominant culture. Strength and perseverance are greatly valued. Many try to conform to these norms by distracting themselves from pain through busyness, entertainment, activity; squashing down or soothing pain with sugar, food, shopping, travel, or sex; or numbing their feelings with drugs, medication, or alcohol. What is not yet so common is to create welcoming, brave, shared spaces where we can acknowledge that we are all affected, all vulnerable at times, all needing support and holding to move through life's losses and challenges. Facing overwhelming issues such as ecological disaster and the biodiversity crisis with others is courageous, effective, and meaningful work.

What follows is a description of some elements that help create a shared grief-tending space.

Tending Grief Together

Tending—to reach out toward, to take care of, to hold. To attend to, to intend.

Grief—a word that can open a doorway for many feelings.

Together—in a shared grief space everyone can take the roles of both the griever and the witness.

The Journey toward Grief

The process of tending grief together follows a sequence of phases. Below I describe the phases of invitation; opening the space; evoking, expressing, and witnessing grief; and integrating the experience and finding a way to return to daily life.

In order to safely hold the space, I always work with a small team of facilitators.

The Invitation

For most of human history, ceremonies have been a birthright, part of the commons, a resource and source of healing, nourishment, and connection for every person in the village. In Western culture, "ceremony" has become a word used with suspicion. Within the grief-tending community throughout the world, a reclaiming process has gained momentum. We reclaim our right and our capacity to create a space together where community can be the container that allows grief to transform. We do this in a way that is not affiliated with any particular religion, faith, or worldview. The communal space that is created extends a welcome to all.

As an insistence on rigid terminology can be excluding, let's call it "a communal space in which transformation might occur." Feel free to find your own language for it.

Opening the Space—Building the Banks of the River

The metaphor of a river can be helpful in understanding the process of releasing grief. Like water, grief flows naturally when the conditions are right. Like water, we can create banks to guide and contain the flow. Strong banks support a strong flow of grief.

The process begins with attending to the conditions that help us to meet grief. The concept of *pendulation*—an oscillatory movement between a place of resource and a place of pain—has been shown to help metabolize strong experiences. This guiding principle underpins the design of the flow and is introduced at the start.

We continue with an introduction to key ideas, and we make group agreements based on care and respect for self, others, the group journey, and the space. This includes confidentiality and respect for different styles, needs, and forms of expression.

We acknowledge that we start with different levels of familiarity with grief. Some people may be close to their grief and impatient to start, while others may feel disconnected from their feelings, perhaps even unsure whether they will reach them at all.

We invite people to connect with what resources and supports them, building a sense of ground and trust.

Evoking, Expressing, Witnessing Grief

In order to create safety, we approach grief gently, finding ways to bring it into awareness. For instance, a method I call "grief soup" invites participants to name a flavor of grief they are carrying. They are encouraged to add it into the mix. This might include the death of a parent, a child excluded from school, the cutting down of the local woodland to build houses, the stress of being unable to pay bills. As the list extends, people are reminded of other griefs they are touched by. This helps to dissolve feelings of anxiety about bringing vulnerability. We are all vulnerable. We all feel things. We are all affected.

As we move toward deeper grief, we remember again what supports us, what is here in the space to anchor us in the present moment, something to touch or hold, an image, a word, a memory, the feeling of a blanket.

Here are two examples of a longer expression of grief:

One is a sharing circle. Those who wish to take a turn come into the center and express whatever is present for them by using words, sounds, silence, tears, or movement, or in some other way. Different objects may be present to support the expression of various feelings. Sometimes people pour water into a vessel or light a candle to acknowledge what they have shared. Each person's sharing is witnessed with kind attention by the circle and acknowledged with a simple phrase at the end: "I hear you," "I see you," "Thank you." At the end of the process we offer the water to the Earth. Some imagine it carrying the intention of what has been shared, perhaps as a blessing to life.

The second example draws on a tradition of the Dagara people, learned from Sobonfu Somé, which I will call a ceremony. At one end of the room we create a place where grief will be expressed, with a cloth laid on a table and decorated with objects. We might call this a grief shrine, a place where the direct expression of grief in all of its facets is expressly invited and allowed. We also create a similar place that helps to resource us, sometimes called a support shrine. We recognize that the word "shrine" may have associations that feel helpful or unhelpful to people, or that may not mean much at all. Once again, people are free to find the words that sit well with their own traditions, culture, and sensitivities. For the purposes of this article I will call it "shrine." Each shrine is created with care, decorated with objects that help evoke its purpose. Some objects are placed by participants.

The ceremony opens with calling in support, reminding us again of what gives life and what holds us steady.

A simple song is repeated, perhaps with a drumbeat. In their own time, participants move between the support shrine and the grief shrine as they wish. More than one person can be at each shrine. Everyone who moves to the grief shrine has someone standing behind them, available to offer physical support if this is asked for and otherwise offering support with their empathic presence. The song and the drumbeat help grief to flow. People move between singing, grieving, and supporting. As grief moves and flows, the sound rises and falls like waves passing through.

The third and vitally important part is to welcome people back into the circle after their time at the grief and support shrines. This is done gently with a kind look or bow; with words, a gentle touch, or hug, offering gratitude for this person's tender heart, their courage, and their work at the shrine.

For some, the experience of being thanked for expressing grief—instead of being told that they are too much—can break their tears open again. This may take them back to the shrine.

I have seen people move enormous amounts of stored-up grief in this process. I have witnessed people finding a way to release feelings they have held in for decades or did not even know they were carrying.

The ceremony ends with an expression of thanks for the support we have called upon or felt was with us. We carry the water from the grief shrine to the land, where it is offered as a gift or a blessing to life. If it feels right, the ceremony can be dedicated to those who have gone before, who could not grieve; those still to come, that they may be free from the burden of our grief; and the web of life, that our expression of grief may be of service to all parts of the living planet.

Integrating and Preparing to Return

After the tender sharing of grief, we allow time and space for the body to settle and integrate the experience. There might be a session of soothing, a shared meal, time outside.

Often there is a feeling of celebration, of deep connection and warmth, and of something important having been accomplished together. People regularly report feeling lightness and joy.

The final processes are designed to facilitate the return to daily life. We acknowledge that there is always more. We reflect on what goes forward from this time and how we might support each other. We ask if there is any intention we want to remember as we leave.

What Is Powerful about Tending Grief, Together?

My distillation of what makes this practice powerful, necessary and transformative:

In this process each person can experience themselves as both vulnerable and capable. Participants may have a sense of being dissolved by grief in one moment, and soon afterward they may stand behind someone to witness their grief and offer them support. The roles of supporter and supported are dissolved. Grief is restored to the commons.

Together we can face the kind of grief that is too overwhelming for individual responses, like global and historical systems of injustice, planetary-scale ecological devastation, or other forms of collective trauma.

Intentional grief spaces allow pain to be metabolized into an energy that supports life. Many feel a new sense of meaning and purpose.

Sharing grief is one of the most powerful ways to build and repair trust and connection. The intensity of grief shared is matched by the intensity of compassion and love that arises when we meet each other in our profound vulnerability.

I believe that grief-tending practices are a missing piece in the endeavor to create a healthy culture. These practices demonstrate what interdependence feels like; thus, they offer an important balance to the cultural overemphasis on individualism, strength, and action.

We need each other to be fully present to all that life brings, the beauty and love and the terror and horror of life. All of it flowing through us all, differently and inseparably.

I am grateful to all my teachers, and to all who are holding and weaving living traditions of tending grief, together.

Bibliography

Introduction

Bauman, Z. (2000). *Liquid modernity.* Polity Press.

Hillman, J. (1995). A psyche the size of the Earth: A psychological foreword. In T. Roszak, M. E. Gomes, & A. D. Kanner (Eds.), *Ecopsychology: Restoring the Earth, healing the mind* (pp. xvii–xxiii). Sierra Club Books.

Macy, J., & Brown, M. Y. (2014). *Coming back to life: The updated guide to the work that reconnects* (Revised ed.). New Society Publishers.

Overview

Solnit, R. (2016). *Hope in the dark: Untold histories, wild possibilities.* Canongate Books.

Transcontextual Reflections on Therapy

Bateson, N. (2016). *Small arcs of larger circles: Framing through other patterns.* Triarchy Press.

Bateson, N. (2022a). New words to hold the invisible world of possibility. Part I: Warm data. *Unpsychology, 8.* https://unpsychology.substack.com/p/new-words-to-hold-the-invisible-world

Bateson, N. (2022b). *Tearing and mending: Transcontextual learning and "healing."* Medium. https://medium.com/@norabateson/tearing-and-mending-e543abf29248

Gilligan, C. (2002). *The birth of pleasure.* Chatto and Windus.

Hillman, J. (1996). *The soul's code: In search of character and calling.* Random Hope.

Hillman, J., & Ventura, M. (1993). *We've had a hundred years of psychotherapy—And the world's getting worse.* Harper.

Le Guin, U. (2017). It doesn't have to be the way it is. In *No time to spare* (pp. 80–85). Houghton Mifflin Harcourt.

Psychotherapy at a Cultural Threshold

Abram, D. (1997). *The spell of the sensuous: Perception and language in a more-than-human world.* Vintage.

Hillman, J., & Ventura, M. (1993). *We've had a hundred years of psychotherapy—And the world's getting worse.* Harper.

Kassouf, S. (2022, February). Thinking catastrophic thoughts: A traumatized sensibility on a hotter planet. *American Journal of Psychoanalysis, 82*(1): 60–79.

Kimbles, S. (2014). *Phantom narratives: The unseen contributions of culture to psyche.* Rowman & Littlefield.

Milstein, T., & Castro-Sotomayor, J. (2020). Ecocultural identity in the Humilocene: An interview with David Abram. https://www.davidabram.org /essays/routledge-handbook-interview

Singer, T., & Kimbles, S. (Eds.). (2004). *The cultural complex: Contemporary Jungian perspectives on psyche and society.* Brunner-Routledge.

Winnicott, D. W. (1971). *Playing and reality.* Penguin.

Why Aren't We Talking about Climate Change? Defenses in the Therapy Room

Bednarek, S. (2019a). Is there a therapy for climate-change anxiety? *Therapy Today, 30*(5): 36–39.

Bednarek, S. (2019b). "This is an emergency"—proposals for a collective response to the climate crisis. *British Gestalt Journal, 28*(2): 4–13.

Benjamin, J. (2004). Beyond doer and done to: An intersubjective view of thirdness. *Psychoanalytic Quarterly, 73*(1): 5–46.

Benjamin, J. (2009). A relational psychoanalysis perspective on the necessity of acknowledging failure in order to restore the facilitating and containing features of the intersubjective relationship (the shared third). *International Journal of Psychoanalysis, 90*(3): 441–450.

Bion, W. R. (1962). *Learning from experience.* Karnac.

Braun, V., & Clarke, V. (2006). Using thematic analysis in psychology. *Qualitative Research in Psychology, 3*(2): 77–101.

Braun, V., & Clarke, V. (2019). Reflecting on reflexive thematic analysis. *Qualitative Research in Sport, Exercise and Health, 11*(4): 589–597.

Braun, V., & Clarke, V. (2021). Can I use TA? Should I use TA? Should I not use TA? Comparing reflexive thematic analysis and other pattern-based qualitative analytic approaches. *Counselling and Psychotherapy Research, 21*(1): 37–47.

Brown, S. (Ed.). (2021). Rewilding therapy: Embracing the environment and breaking the silence on climate change [Special issue]. *Therapy Today, 32*(9). www.bacp.co.uk/bacp-journals/therapy-today/2021/november-2021/

Carbon conversations. (n.d.). Carbon conversations. Retrieved June 30, 2022, from www.carbonconversations.co.uk/p/about.html

Cartwright, D. (2004). The psychoanalytic research interview: Preliminary suggestions. *Journal of the American Psychoanalytic Association, 52*(1): 209–242.

Clarke, S., & Hogget, P. (2009). Researching beneath the surface: A psycho-social approach to research, practice and method. In S. Clarke & P. Hogget (Eds.), *Researching beneath the surface: Psycho-social research methods in practice* (pp. 17–49). Karnac.

Frosh, S., Phoenix, A., & Pattman, R. (2003). Taking a stand: Using psychoanalysis to explore the positioning of subjects in discourse. *British Journal of Social Psychology, 42*(1): 39–53.

Hamilton, J. (2019). Emotions, reflexivity and the long haul: What we do about how we feel about climate change. In P. Hoggett (Ed.), *Climate psychology: On indifference to disaster* (pp. 153–175). Springer Nature.

Haraway, D. J. (2016). *Staying with the trouble: Making kin in the Chthulucene.* Duke University Press.

Hickman, C., Marks, E., Pihkala, P., Clayton, S., Lewandowski, E., Mayall, E., Wray, B., Mellor, C., & van Susteren, L. (2021). Climate anxiety in children and young people and their beliefs about government responses to climate change: A global survey. *The Lancet Planetary Health, 5*(12): 863–873.

Hollway, W., & Jefferson, T. (2000). *Doing qualitative research differently: Free association, narrative and the interview method.* Sage.

Hollway, W., & Jefferson, T. (2013). *Doing qualitative research differently* (2nd ed.). Sage.

IPCC (Intergovernmental Panel on Climate Change). (2022). *Climate change 2022: Impacts, adaptation and vulnerability. Summary for policy makers.* https://report.ipcc.ch/ar6wg2/pdf/IPCC_AR6_WGII_SummaryForPolicy makers.pdf

Jackson, C. (2021). The big interview: Nick Totton. *Therapy Today, 32*(9): 26–30.

Lertzman, R. (2015). *Environmental melancholia: Psychoanalytic dimensions of engagement.* Routledge.

Macy, J., & Johnstone, C. (2012). *Active hope: How to face the mess we're in without going crazy.* New World Library.

Macy, J., & Young Brown, M. (1998). *Coming back to life: Practices to reconnect our lives.* New Society Publishing.

Menzies Lyth, I. (1960). Social systems as a defence against anxiety. *Human Relations, 13*: 95–121.

Patton, M. Q. (2002). *Qualitative research & evaluative methods* (3rd ed.). Sage.

RCP (Royal College of Psychiatrists). (2021). *Position statement PS03/21: Our planet's climate and ecological emergency.* www.rcpsych.ac.uk/docs/default -source/improving-care/better-mh-policy/position-statements/position -statement-ps03-21-climate-and-ecological-emergencies-2021.pdf

Rust, M.-J. (2020). *Towards an ecopsychotherapy.* Confer Books.

Samuels, A. (2001). *Politics on the couch: Citizenship and the internal life.* Other Press.

Samuels, A. (2006). Politics on the couch? Psychotherapy and society—Some possibilities and some limitations. In N. Totton (Ed.), *The politics of psychotherapy: New perspectives* (pp. 3–16). Open University Press/McGraw-Hill Education.

Siegel, D. (1999). *The developing mind: How relationships and the brain interact to shape who we are.* Guilford Press.

UKCP (UK Council for Psychotherapy). (2016). *The Psychotherapist, 63.* Summer.

Wainwright, T., & Mitchell, A. (Eds.). (2020). Psychology and the climate and environmental crisis [Special issue]. *Clinical Psychology Forum, 332.* https:// explore.bps.org.uk/content/bpscpf/1/332

Weintrobe, S. (2013). Introduction. In S. Weintrobe (Ed.), *Engaging with climate change: Psychoanalytic and interdisciplinary perspectives* (pp. 1–15). Routledge.

Weintrobe, S. (2021). *Psychological roots of the climate crisis: Neoliberal exceptionalism and the culture of uncare.* Bloomsbury Academic.

Wengraf, T. (2001). *Qualitative research interviewing.* Sage.

Winnicott, D. W. (1971). *Playing and reality.* Routledge.

Zerubavel, E. (2006). *The elephant in the room: Silence and denial in everyday life.* Oxford University Press.

Frozen in Trauma on a Warming Planet:
A Relational Reckoning with Climate Distress

Ahmed, N. (2020, June 5). White supremacism and the Earth system. Medium. https://medium.com/insurge-intelligence/white-supremacism-and-the -earth-system-fa14e0ea6147

Baldwin, J. (1962, November 17). Letter from a region in my mind. *New Yorker.* www.newyorker.com/magazine/1962/11/17/letter-from-a-region -in-my-mind

Baudon, P., & Jachens, L. (2021). A scoping review of interventions for the treatment of eco-anxiety. *International Journal of Environmental Research in Public Health, 18:* 9636.

Bednarek, S. (2019). "This is an emergency"—Proposals for a collective response to the climate crisis. *British Gestalt Journal, 28*(2): 4–13.

Bednarek, S. (2021). A time of derangement. *Resurgence & Ecologist, 327:* 25–27.

Craps, S. (2014). Beyond Eurocentrism: Trauma theory in the Golden Age. In G. Buelens, S. Durrant, & R. Eaglestone (Eds.), *The future of trauma theory: Contemporary literary and cultural criticism* (pp. 45–61). Routledge.

Doherty, T. & Clayton, S. (2011). The psychological impacts of global climate change. *American Psychologist, 66*(4): 265–276.

Doppelt, B. (2016). *Transformational resilience: How building human resilience to climate disruption can safeguard society and increase well-being.* Routledge.

Dunlop, D. (2015). Composting pain. https://wisebirds.org/2015/04/26 /hello-world/

Eagle, G., & Kaminer, D. (2013). Continuous traumatic stress: Expanding the lexicon of traumatic stress. *Peace and Conflict: Journal of Peace Psychology, 19*(2): 85–99.

Fisher, A. (2012). What is ecopsychology: A radical view. In P. H. Kahn & P. H. Hasbach (Eds.), *Ecopsychology: Science, totems and the technological species* (pp. 78–114). MIT Press.

Fonagy, P., & Allison, E. (2014). The role of mentalizing and epistemic trust in the therapeutic relationship. *Psychotherapy, 10*(1037): a0036505.

Goodreads (2008). Interview with Paulo Coelho. www.goodreads.com/interviews /show/2.Paulo_Coelho

Goodreads (n.d.). *Ryunosuke Satoro > Quotes > Quotable Quote.* Retrieved February 1, 2023, from www.goodreads.com/quotes/479992-individually -we-are-one-drop-together-we-are-an-ocean

Hanh, T. (1993). *Interbeing: Fourteen guidelines for engaged Buddhism.* Parallax Press.

Haseley, D. (2019). Climate change: Clinical considerations. *International Journal of Applied Psychoanalytic Studies, 16*(2): 109–115.

Heglar, M. A. (2018). Climate change isn't the first existential threat. Medium. https://zora.medium.com/sorry-yall-but-climate-change-ain-t-the-first-existential-threat-b3c999267aa0

Hoggett, P., & Randall, R. (2018). Engaging with climate change: Comparing the cultures of science and activism. *Environmental Values, 27*: 223–243.

Hopenwasser, K. (2018). The rhythm of resilience: A deep ecology of entangled relationality. In J. Salberg & S. Grand (Eds.), *Wounds of history: Repair and resilience in the trans-generational transmission of trauma* (pp. 60–76). Routledge.

Hurston, Z. N. (2009). In Tram Nguyen (Ed.), *Language is a place of struggle: Great quotes by people of color* (p. 96). Beacon Press.

IPCC (Intergovernmental Panel on Climate Change). (2018). *Special report: Global warming of 1.5°C.* www.ipcc.ch/sr15/

Kaplan, E. A. (2015). *Climate trauma: Foreseeing the future in dystopian film and fiction.* Rutgers University Press.

Kassouf, S. (2022). Thinking catastrophic thoughts: A traumatized sensibility on a hotter planet. *American Journal of Psychoanalysis, 82*: 60–79.

Kimmerer, R. (2013). *Braiding sweetgrass: Indigenous wisdom, scientific knowledge and the teachings of plants.* Milkweed Editions.

Lertzman, R. (2013). The myth of apathy: Psychoanalytic explorations of environmental subjectivity. In S. Weintrobe (Ed.), *Climate change: Psychoanalytic and interdisciplinary perspectives* (pp. 117–133). Routledge.

Lewis, J., Haase, E., & Trope, A. (2020). Climate dialectics in psychotherapy: Holding open the space between abyss and advance. *Psychodynamic Psychiatry, 48*(3): 271–294.

Lifton, R. J. (2017). *The climate swerve: Reflections on mind, hope, and survival.* The New Press.

Marshall, G. (2014). *Don't even think about it: Why our brains are wired to ignore climate change.* Bloomsbury.

Menakem, R. (2019). *My grandmother's hands: Racialized trauma and the pathway to mending our hearts and bodies.* Central Recovery Press.

Mitchell, S. (2020). Indigenous prophecy and Mother Earth. In A. Johnson & K. Wilkinson (Eds.), *All we can save: Truth, courage and solutions for the climate crisis* (pp.16–28). Random House.

Morton, T. (2013). *Hyperobjects: Philosophy and ecology after the end of the world.* University of Minnesota Press.

Orange, D. (2017). *Climate crisis, psychoanalysis and radical ethics.* Routledge.

Randall, R. (2005). A new climate for psychotherapy? *Psychotherapy and Politics International, 3*: 165–179.

Salberg, J. (2015). The texture of traumatic attachment: Presence and ghostly absence in transgenerational transmission. *Psychoanalytic Quarterly, 84*(1): 21–46.

Searles, H. (1960). *The nonhuman environment in normal development and in schizophrenia.* International Universities Press.

Siegel, D. (1999). *The developing mind: How relationships and the brain interact to shape who we are.* Guilford Press.

Tedeschi, R. G., & Calhoun, L. G. (1996). The Posttraumatic Growth Inventory: Measuring the positive legacy of trauma. *Journal of Traumatic Stress, 9*(3): 455–472.

Terr, L. (1990). *Too scared to cry: Psychic trauma in childhood.* Basic Books.

Weintrobe, S. (2020). *Working through our feelings about the climate crisis* [Conference session]. The Climate Emergency: Psychoanalytic Perspectives online conference, May 23, 2020.

Weintrobe, S. (2021). *Psychological roots of the climate crisis: Neoliberal exceptionalism and the culture of uncare.* Bloomsbury Academic.

Woodbury, Z. (2019). Toward a new taxonomy of trauma. *Ecopsychology, 11*(1): 1–8.

Woods, C. (1998). *Development arrested: The blues and plantation power in the Mississippi Delta.* Verso Books.

Wordsworth, W., & Coleridge, S. (2011). *Lyrical Ballads 1798: A Critical Edition by William Wordsworth and Samuel Taylor Coleridge.* Clemson University Digital Press. https://tigerprints.clemson.edu/cgi/viewcontent.cgi?article=1005&context=cudp_bibliography

Climate Change and Thirst

American Psychological Association, APA Task Force on Climate Change. (2022). *Addressing the climate crisis: An action plan for psychologists. Report of the APA Task Force on Climate Change.* www.apa.org/science/about/publications/ climate-crisis-action-plan.pdf

Hogan, C. (Host). (2021, September 29). Will we run out of WATER? We need to talk about eco-anxiety [Audio podcast episode]. In *Force of nature.* https://tinyurl.com/3x963d85

Mekonnen, M., & Hoekstra, A. (2016). Four billion people facing severe water scarcity. *Science Advances, 2*(2): e1500323. www.ncbi.nlm.nih.gov/pmc/articles/PMC4758739/

Olander, E. (2022, 28 February). South Africa's Limpopo province approves $10 billion Chinese-finance coal project . . . but there's just one very small problem. https://chinaglobalsouth.com/2022/02/28/south-africas-limpopo-province-approves-10-billion-chinese-finance-coal-project-but-theres-just-one-very-small-problem/

United Nations (n.d.). *COP26: Together for our planet.* www.un.org/en/climatechange/cop26

The Visibly Invisible Shadow: Decolonizing Work in Environmental Movements

Bateson, N. (2021). What is submerging? Medium. https://norabateson.medium.com/what-is-submerging-ad12df016cde

Bendell, J. (2020, June 28). The collapse of ideology and the end of escape. https://jembendell.com/2020/06/28/the-collapse-of-ideology-and -the-end-of-escape

DAF (Deep Adaptation Forum). (n.d.). Embodying and enabling loving responses to our predicament. Retrieved November 19, 2022, from www .deepadaptation.info

DiAngelo, R. (2016). *What does it mean to be white?* Peter Lang.

Henderson, J. A. (2019). Learning to teach climate change as if power matters. *Environmental Education Research*, 25: 987–990.

hooks, b. (2014). *Sisters of the yam: Black women and self-recovery*. Routledge.

Zweig, C., & Abrams, J. (1991). *Meeting the shadow: The hidden power of the dark side of human nature*. Penguin.

Decolonizing Psychotherapy

Asmus, Katie & Bryson, Hāweatea Holly (2020). *Ceremony & rites of passage global training* (n.d.). Retrieved September 26, 2023, from www.ritesof passagetraining.com

Atkinson, J. (2002). *Trauma trails, recreating songlines*. Spinifex Press.

Fanon, F. (1963). *The wretched of the Earth*. Grove Press.

Haeyoung, J. (2013). Consideration of Indigenous ethos in psychotherapeutic practices: *Pungryu* and Korean psychotherapy. *Asia Pacific Journal of Counselling and Psychotherapy*, 10: 2.

Hall, E. T. (1976). *Source beyond culture*. Indiana Department of Education.

ICT (Indigenous Corporate Training). (2020). Indigenous title and the doctrine of discovery. www.ictinc.ca/blog/indigenous-title-and-the-doctrine-of-discovery

Just Reinvest NSW (n.d.). Retrieved September 26, 2023, from www.justreinvest .org.au

Lange, R. (1999). *May the people live: A history of Maori health development 1900–1920*. Auckland University Press.

Liberman, A. (2007). *An analytic dictionary of English etymology: An introduction*. University of Minnesota Press.

Meyer, J., & Land, R. (2003). *Threshold concepts and troublesome knowledge: Linkages to ways of thinking and practising within the disciplines*. University of Edinburgh.

Miller, R. J. (2017). The Doctrine of Discovery: The international law of colonialism. *UCLA Indigenous Peoples' Journal of Law, Culture, and Resistance*. http://dx.doi.org/10.2139/ssrn.3541299

Neilson, M. (2020, September 13). Te wiki o te reo Māori: Beaten for speaking their native tongue, and the generations that suffered. *New Zealand Herald*. https://www.nzherald.co.nz/nz/te-wiki-o-te-reo-maori-beaten -for-speaking-their-native-tongue-and-the-generations-that-suffered /F7G6XCM62QAHTYVSRVOCRKAUYI/

Noon, K., & De Napoli, K. (2022). *Astronomy: Sky country*. Thames & Hudson.

Petchkovsky, L., San Roque, C., & Beskow, M. (2003). Jung and the dreaming: Analytical psychology's encounters with Aboriginal culture. *Transcultural Psychiatry, 40*(2): 208–238.

San Roque, C. (2012). Aranke or in the long line: Reflections on the 2012/2011 Sigmund Freud Award for Psychotherapy and the lineage of traditional Indigenous therapy in Australia. *Psychotherapy and Politics International, 10*(2): 93–104.

Stephens, M. (2001). *A return to Tohunga Suppression Act 1907. Victoria University of Wellington Law Review, 32*(2): 437–462.

Tuck, E., & Wayne Yang, K. (2012). Decolonization is not a metaphor. *Decolonization: Indigeneity, Education & Society, 1*(1): 1–40.

Van Gennep, A. (1909). *The rites of passage.* University of Chicago Press.

Webster, P. (1979). *Rua and the Maori millennium.* Victoria University Press.

Anthropocentrism, Animism, and the Anthropocene: Decentering the Human in Psychology

Adams, M., Ormrod, J., & Smith, S. (2023). Notes from a field: A qualitative exploration of human-animal relations in a volunteer shepherding project. *Qualitative Research, 23*(1): 163–172.

Bastian, B., and Loughnan, S. (2017). Resolving the meat-paradox: A motivational account of morally troublesome behavior and its maintenance. *Personality and Social Psychology Review, 21*(3): 278–299.

Bird-Rose, D. (2017). Connectivity thinking, animism, and the pursuit of liveliness. *Educational Theory, 67*(4): 491–508.

Bolman, B. (2019). Parroting patriots: Interspecies trauma and becoming-well -together. *Medical Humanities, 45*(3): 305–312.

Bradshaw, G. A. (2010). You see me, but do you hear me? The science and sensibility of trans-species dialogue. *Feminism & Psychology, 20*(3): 407–419.

Country, B., Wright, S., Suchet-Pearson, S., Lloyd, K., Burarrwanga, L., Ganambarr, R., Ganambarr-Stubbs, M., Ganambarr, B., & Maymuru, D. (2015). Working with and learning from country: Decentering human authority. *Cultural Geographies, 22*(2): 269–283.

Fields, T. R. (2020). Trees in early Irish law and lore: Respect for other-than -human life in Europe's history. *Ecopsychology, 12*(2): 130–137.

Fisk, A. (2017). Appropriating, romanticizing and reimagining: Pagan engagements with indigenous animism. In *Cosmopolitanism, Nationalism, and Modern Paganism* (pp. 21–42). Palgrave Macmillan.

Ghosh, A. (2016). *The great derangement: Climate change and the unthinkable.* Penguin.

Gibson, P. R. (2019). Waking up to the environmental crises. *Ecopsychology, 11*(2): 67–77.

Gorman, R. (2019). What's in it for the animals? Symbiotically considering "therapeutic" human–animal relations within spaces and practices of care farming. *Medical Humanities, 45*(3): 313–325.

Haraway, D. (2016). *Staying with the trouble.* Duke University Press.

Harvey, G. (2006). *Animism: Respecting the living world.* Columbia University Press.

Harvey, G. (2019). Animism and ecology: Participating in the world community. *The Ecological Citizen, 3*(1): 79–84.

Hayward, T. (1997). Anthropocentrism: A misunderstood problem. *Environmental Values, 6*(1): 49–63.

Hruschka, R. (2018, November 16). You can't characterize human nature if studies overlook 85 percent of people on Earth. *The Conversation.* https://theconversation.com/you-cant-characterize-human-nature-if-studies-overlook-85-percent-of-people-on-earth-106670

Kessi, S., Suffla, S., & Seedat, M. (2022). *Decolonial enactments in community psychology.* Springer.

Kimmerer, R. W. (2017). Learning the grammar of animacy. *Anthropology of Consciousness, 28*(2): 128–134.

Lang, A. (1899). *Myth, ritual and religion, vol. 2.* Longmans, Green.

Lien, M. E., & Pálsson, G. (2021). Ethnography beyond the human: The "other-than-human" in ethnographic work. *Ethnos, 86*(1): 1–20.

Pedersen, H. (2021). Education, anthropocentrism, and interspecies sustainability: Confronting institutional anxieties in omnicidal times. *Ethics and Education, 16*(2): 164–177.

Peralta, J. M., and Fine, A. H. (2021). The welfarist and the psychologist: Finding common ground in our interactions with therapy animals. In *The welfare of animals in animal-assisted interventions* (pp. 265–284). Springer.

Plumwood, V. (2007). A review of Deborah Bird Rose's *Reports from a wild country: Ethics of decolonisation. Australian Humanities Review,* 42: 1–4.

Plumwood, V. (2009). Nature in the active voice. *Australian Humanities Review,* 46: 1–13.

Rautio, P., Hohti, R., Leinonen, R. M., & Tammi, T. (2017). Reconfiguring urban environmental education with "shitgull" and a "shop." *Environmental Education Research, 23*(10): 1379–1390.

Reid, J., & Rout, M. (2016). Getting to know your food: The insights of Indigenous thinking in food provenance. *Agriculture and Human Values, 33*(2): 427–438.

Robbins, J. (2018, April 26). Native knowledge: What ecologists are learning from Indigenous people. *Yale Environment 360.* https://e360.yale.edu/features/native-knowledge-what-ecologists-are-learning-from-indigenous-people

Robinson, A. S. (2019). Finding healing through animal companionship in Japanese animal cafés. *Medical Humanities, 45*(2): 190–198.

Rothberger, H., & Rosenfeld, D. L. (2021). Meat-related cognitive dissonance: The social psychology of eating animals. *Social and Personality Psychology Compass, 15*(5): e12592.

Rountree, K. (2012). Neo-paganism, animism, and kinship with nature. *Journal of Contemporary Religion, 27*(2): 305–320.

Serpell, J. A. (2010). Animal-assisted interventions in historical perspective. In A. H. Fine (Ed.), *Handbook on animal-assisted therapy* (pp. 17–32). Academic Press.

Shapiro, K. (2020). Human–animal studies: Remembering the past, celebrating the present, troubling the future. *Society & Animals, 28*(7): 797–833.

Siebert, C. (2016, January 28). What does a parrot know about PTSD? *New York Times.* www.nytimes.com/2016/01/31/magazine/what-does-a -parrot-know-about-ptsd.html

Stacey T. (2021). Toying with animism: How learning to play might help us get serious about the environment. *Nature + Culture, 16*(3): 83–109.

Taylor, N., Fraser, H., & Riggs, D. W. (2020). Companion-animal-inclusive domestic violence practice: Implications for service delivery and social work. *Aotearoa New Zealand Social Work, 32*(4): 26–39.

Todd, Z. (2015). Indigenizing the Anthropocene. In H. Davis and E. Turpin (Eds.), *Art in the Anthropocene: Encounters among aesthetics, politics, environments and epistemologies* (pp. 241–254). Open Humanities Press.

Van Dooren, T., Kirksey, E., & Münster, U. (2016). Multispecies studies: Cultivating arts of attentiveness. *Environmental Humanities, 8*(1): 1–23.

Washington, H., Piccolo, J., Gomez-Baggethun, E., Kopnina, H., & Alberro, H. (2021). The trouble with anthropocentric hubris, with examples from conservation. *Conservation, 1*(4): 285–298.

Watts, V. (2013). Indigenous place-thought and agency amongst humans and non humans (First Woman and Sky Woman go on a European world tour!). *Decolonization: Indigeneity, Education & Society, 2*(1): 20–34.

Westerlaken, M. (2021). What is the opposite of speciesism? On relational care ethics and illustrating multi-species-isms. *International Journal of Sociology and Social Policy, 41*(3/4): 522–540.

Whyte, K. (2017). Indigenous climate change studies: Indigenizing futures, decolonizing the Anthropocene. *English Language Notes, 55*(1): 153–162.

Psychotherapy, Anthropocentrism, and the Family of Things

Barrow, G., & Marshall, H. (Eds.). (2023). EcoTA [Special issue]. *Transactional Analysis Journal, 53*(1).

Bateson, G. (1972). *Steps to an ecology of mind.* Chandler Publishing.

Bednarek, S. (2019). How wide is the field?: Gestalt therapy, capitalism and the natural world. *Gestalt Journal of Australia and New Zealand, 15*(2): 18–42.

Berry, T. (1987). The dream of the Earth: Our way into the future. *CrossCurrents, 37*(2/3): 200–215. www.jstor.org/stable/24459049

Berry, T. (2013). The great work of the new millennium. *The NAMTA Journal, 38*(1): 249–256.

Bogost, I. (2012). *Alien phenomenology, or what it's like to be a thing.* University of Minnesota Press.

Braidotti, R. (2013). *The posthuman.* Polity Press.

Carson, R. (1962). *Silent spring.* Houghton Mifflin.

Chidiac, M.-A., & Denham-Vaughan, S. (2007). The process of presence: Energetic availability and fluid responsiveness. *British Gestalt Journal, 16*(1): 9–19.

Ferrer, J. (2002). *Revisioning transpersonal theory: A participatory vision of human spirituality.* State University of New York Press.

Freud, S. (2002). *Civilisation and its discontents* (D. McLintock, Trans.). Penguin Books.

Garrison Institute. (2015, Feb. 24). *Dan Siegel on neurobiology and resilience* [Video]. YouTube. https://youtu.be/Zriw-jShjzY?t=5

Haraway, D. (1991). A cyborg manifesto: Science, technology, and socialist-feminism in the late twentieth century. In *Simians, cyborgs and women: The reinvention of nature* (pp. 149–181). Routledge.

Harman, G. (2005). *Guerrilla metaphysics: Phenomenology and the carpentry of things.* Open Court.

Haywood, T. (1997). Anthropocentrism: A misunderstood problem. *Environmental Values, 6*(1): 49–63. www.jstor.org/stable/30301484

Juniper, T. (2022). *What are they worth?* RSPCA. www.rspca.org.uk/whatwedo/latest/essays/whataretheyworth

Keys, S. (Ed.). (2013). Ecology and person-centered and experiential psychotherapies [Special issue]. *Person-Centered & Experiential Psychotherapies, 12*(4).

Latour, B. (2005). *Reassembling the social: An introduction to actor–network theory.* Oxford University Press.

Næss, A. (1972). Shallow and the deep: Long-range ecology movements: A summary. *Inquiry, 16*(1): 95–100. https://doi.org/10.1080/00201747308601682

Næss, A., & Sessions, G. (1986). *The basic principles of deep ecology.* https://theanarchistlibrary.org/library/arne-naess-and-george-sessions-basic-principles-of-deep-ecology

Oliver, M. (1986). *Dream Work.* Atlantic Monthly Press.

Perls, F., Hefferline, R., & Goodman, P. (1951). *Gestalt therapy: Excitement and growth in the human personality.* Julian Press.

Plotkin, B. (2008). *Nature and the human soul: Cultivating wholeness and community in a fragmented world.* New World Library.

Plotkin, B. (2021). *The journey of soul initiation: A field guide for visionaries, evolutionaries and revolutionaries.* New World Library.

Prendergast, M. (2009). "Poem is what?" Poetic inquiry in qualitative social science research. In M. Prendergast, C. Leggo, & P. Sameshima (Eds.), *Poetic inquiry: Vibrant voices in the social sciences* (pp. 357–362). Brill.

Prescott, S. L., Logan, A. C., Bristow, J., Rozzi, R., Moodie, R., Redvers, N., Haatea, T., Warber, S., Poland, B., Hancock, T., & Berman, B. (2022). Exiting the Anthropocene: Achieving personal and planetary health in the 21st century. *Allergy, 77*(12): 1–15. https://doi.org/10.1111/all.15419

Price-Robertson, R. (2020). Beyond phenomena: Facing the reality of the social world. *Gestalt Journal of Australia and New Zealand, 16*(1): 19–40.

Roszak, T. (2001). *The voice of the Earth: An exploration of ecopsychology.* Red Wheel/Weiser.

Roszak, T. E., Gomes, M. E., & Kanner, A. D. (1995). *Ecopsychology: Restoring the Earth, healing the mind.* Sierra Club Books.

Schmid, P. F. (2006). The challenge of the other: Towards dialogical person-centered psychotherapy and counseling. *Person-Centered & Experiential Psychotherapies, 5*(4): 240–254. https://doi.org/10.1080/14779757.2006.9688416

Shepard, P. (1969). *The subversive science: Essays toward an ecology of man.* Houghton Mifflin.

Simard, S. (2021). *Finding the mother tree: Discovering the wisdom of the forest.* Allen Lane.

Skelding, M. (2020). *On equilibrium—People, patterns and psychosphere.* www .academia.edu/48904006/On_Equilibrium_People_Patterns_and _Psychosphere

Thunberg, G. (2019, Jan. 25). *Our house is on fire* [Address]. World Economic Forum, Davos, Switzerland.

Weber, A. (2016). *The biology of wonder: Aliveness, feeling and the metamorphosis of science.* New Society Publishers.

Weber, A. (2017). *Matter and desire: An erotic ecology.* Chelsea Green Publishing.

Psychotherapy as Sumbiography: Dissociation in the Anthropocene and Association in the Symbiocene

Albrecht, G. A. (2012, Sept. 2). Tierratrauma. *Healthearth.* http://healthearth .blogspot.com.au/2012/09/tierratrauma.html?m=0

Albrecht, G. A. (2017, July 27). Ecoagnosy. *Psychoterratica.* https://glennaalbrecht .wordpress.com/2017/07/27/ecoagnosy/

Albrecht, G. A. [glennalbrecht]. (2018, April 16). *Of street trees and solastalgia* [Comment]. Sheffield Tree Action Groups. https://savesheffieldtrees.org.uk /2018/04/16/of-street-trees-and-solastalgia-joanna-dobson/#comments

Albrecht, G. A. (2019). *Earth emotions: New words for a new world.* Cornell University Press.

Albrecht, G. A. (2020a). Negating solastalgia: An emotional revolution from the Anthropocene to the Symbiocene. *American Imago, 77*(1): 9–30.

Albrecht, G. A. (2020b). One hundred years of sumbiotude: Resisting the extinction of emotions. *Griffith Review, 68:* 159–177.

Albrecht, G. A. (2021). Meuacide. *Psychoterratica.* https://glennaalbrecht .wordpress.com/2020/02/02/meuacide-the-extinction-of-emotions/

Albrecht, G. A., Sartore, G.-M., Connor, L., Higginbotham, N., Freeman, S., Kelly, B., Stain, H., Tonna, A., & Pollard, G. (2007). Solastalgia: The distress caused by environmental change. *Australasian Psychiatry, 15* (Special supplement): 95–98.

Arndt, L. (2021). Student sumbiographies. https://issuu.com/laurenarndt/docs /sumbiographies4

Beale, B. (2007). *If trees could speak: Stories of Australia's greatest trees.* Allen and Unwin.

Burin, M. (2018). People from all over the world are sending emails to Melbourne's trees. ABC News. www.abc.net.au/news/2018-12-12/people-are -emailing-trees/10468964?nw=0&r=HtmlFragment

Delalonde, M. (2020). Gen (S). https://www.generationsymbiocene.gr/

Holmes, K., Martin, S. K., & Mirmohamadi, K. (2008). *Reading the garden: The settlement of Australia.* Melbourne University Press.

Johnson, T. (2018). *Radical joy for hard times: Finding meaning and making beauty in Earth's broken places.* North Atlantic Books.

Kahn, P. (1999). *The human relationship with nature: Development and culture.* MIT Press.

Kahn, P. H., Jr., Severson, R., & Ruckert, J. (2009). The human relation with nature and technological nature. *Current Directions in Psychological Science, 18*(1): 37–42. https://depts.washington.edu/hints/publications/Human _Relation_Technological_Nature.pdf

Kellert, S. R., & Wilson, E. O. (Eds.). (1993). *The biophilia hypothesis.* Island Press.

Le Roux, G. (2016). Transforming representations of marine pollution. For a new understanding of the artistic qualities and social values of ghost nets. *Anthrovision, (4)*1. https://doi.org/10.4000/anthrovision.2221

Lertzman, R. (2008, June 19). The myth of apathy. *The Ecologist.* https:// theecologist.org/2008/jun/19/myth-apathy

Louv, R. (2008). *Last child in the woods: Saving our children from nature-deficit disorder.* Algonquin Books.

Louv, R. (2011). *The nature principle: Human restoration and the end of nature-deficit disorder.* Algonquin Books.

Pretty, J. (2017). Manifesto for the green mind. *Resurgence and Ecologist, 301:* 18–23.

Pyle, R. M. (1993). *The thunder tree: Lessons from an urban wildland.* Houghton Mifflin.

Sobel, D. (1996). *Beyond ecophobia: Reclaiming the heart in nature education.* Orion Society.

Young, K. (2004). Daluk rangers. ABC Earthbeat. www.abc.net.au/radionational /programs/archived/earthbeat/daluk-rangers/3640276

Yunkaporta, T. (2019). *Sand talk: How Indigenous thinking can save the world.* Text Publishing.

Climate Change, Fragmentation, and Collective Trauma: Bridging the Divided Stories We Live By

Abram, D. (1997). *The spell of the sensuous: Perception and language in a more-than-human world.* Vintage.

Arendt, H. (2006). *Eichmann in Jerusalem: A report on the banality of evil.* Penguin.

Bateson, G. (1972). *Steps to an ecology of mind: Collected essays in anthropology, psychiatry, evolution, and epistemology.* University of Chicago Press.

Bauman, Z. (2000). *Liquid modernity.* Cambridge: Polity Press.

Bednarek, S. (2018). How wide is the field? Gestalt psychotherapy, capitalism and the natural world. *British Gestalt Journal, 27*(2): 8–17.

Bednarek, S. (2019). "This is an emergency"—Proposals for a collective response to the climate crisis. *British Gestalt Journal, 28*(2): 6–15.

Berger, S., Kouzarides, T., Shiekhattar, R., & Shilatifard, A. (2009). An operational definition of epigenetics. *Genes & Development, 23*(7): 781–783.

Campbell, J. (1949). *The hero with a thousand faces.* Princeton University Press.

Capra, F. (1982). *The turning point: Science, society and the rising culture.* Bantam Books.

Capra, F., & Luisi, P. L. (2014). *The systems view of life: A unifying vision.* Cambridge University Press.

Confino, J. (2015, April 21). Beyond capitalism and socialism: Could a new economic approach save the planet? *The Guardian.* www.theguardian.com /sustainable-business/2015/apr/21/regenerative-economyholism-economy -climate-change-inequality

Erskine, R. (1994). Shame and self-righteousness: Transactional analysis perspectives and clinical interventions. *Transactional Analysis Journal, 24*(2): 86–102.

Fischer, J. (2017). *Healing the fragmented selves of trauma survivors: Overcoming internal self-alienation.* Routledge.

Forner, C. (2017). *Dissociation, mindfulness and creative meditations: Trauma-informed practices to facilitate growth.* Routledge

Glendenning, C. (1994). *My name is Chellis and I'm in recovery from Western civilization.* Shambhala.

Gretton, D. (2019). *I you we them: Journeys beyond evil: The desk killer in history and today.* William Heinemann.

Harvard Medical School. (2019, Dec. 9). *Healing collective trauma* [Video]. YouTube. www.youtube.com/watch?v=mExBoPftp8I

Higgins, P. (2010). *Eradicating ccocide: Laws and governance to prevent the destruction of our planet.* Shepheard-Walwyn.

Hillman, J. (1995). A psyche the size of the earth: A psychological foreword. In T. Roszak, M. E. Gomes, & A. D. Kanner (Eds.), *Ecopsychology: Restoring the Earth, healing the mind* (pp. xiii–xvi). Sierra Club Books.

Hillman, J. (1996). *The soul's code: In search of character and calling.* Random House.

IPCC (Intergovernmental Panel on Climate Change). (2018). *Special report: Global warming of 1.5°C.* www.ipcc.ch/sr15/

Jung, C. G. (1980). *Psychology and alchemy* (2nd ed.). Routledge.

Jung, C. G., & Jaffe, A. (1995). *Memories, Dreams, Reflections.* Fontana Press.

Kremer, J. (1998). The shadow of evolutionary thinking. In D. Rothberg & S. Kelly (Eds.), *Ken Wilber in dialogue* (pp. 237–258). Quest.

Lertzman, R. (2015). *Environmental melancholia: Psychoanalytic dimensions of engagement.* Routledge.

Levine, P. (1997). *Waking the tiger: Healing trauma.* North Atlantic Books.

Mack, J. E. (1995). The politics of species arrogance. In T. Roszak (Ed.), *Ecopsychology: Restoring the Earth, healing the mind* (pp. 279–288). Sierra Club Books.

McGilchrist, I. (2010a). *The master and his emissary: The divided brain and the making of the Western world.* Yale University Press.

McGilchrist, I. (2010b). Recapturing the whole: Brain hemispheres and the renewal of culture. In D. Lorimer and O. Robinson (Eds.), *A new renaissance: Transforming science, spirit and society* (pp. 61–71). Flioris.

McGrane, B. (1989). *Beyond anthropology*. Columbia.

Rohr, R. (2012). *Falling upward: A spirituality for the two halves of life*. Spec.

Rothchild, B. (2000). *The body remembers: The psychophysiology of trauma and trauma treatment*. W. W. Norton.

Sereny, G. (1974). *Into that darkness: From mercy killing to mass murder*. Pimlico.

Short, A. K., Fennell, K. A., Perreau, V. M., Fox, A., O'Bryan, M. K., & Kim, J. H. (2016). Elevated paternal glucocorticoid exposure alters the small noncoding RNA profile in sperm and modifies anxiety and depressive phenotypes in the offspring. *Translational Psychiatry, 6*(6): e837.

Siegel, D. (1999). *The developing mind*. Guilford.

Van der Hart, O., Nijenhus, E., & Steele, K. (2006). *The haunted self: Structural dissociation and the treatment of chronic traumatization*. W. W. Norton.

Van der Kolk, B. (2014). *The body keeps the score: Mind, brain, and body in transformation of trauma*. Penguin Random House.

Weintrobe, S. (Ed.). (2013). *Engaging with climate change: Psychoanalytic and interdisciplinary perspectives*. Routledge.

Weller, F. (2015). *The wild edge of sorrow: Rituals of renewal and the sacred work of grief*. North Atlantic Books.

Woodbury, Z. (2019). Climate trauma: Toward a new taxonomy of trauma. *Ecopsychology, 11*(1): 1–8.

World Bank. (2018, March 19). *Climate change could force over 140 million to migrate within countries by 2050: World Bank report*. www.worldbank.org /en/news/press-release/2018/03/19/climate-change-could-force-over -140-million-to-migrate-within-countries-by-2050-world-bank-report

Four Lectures for the Kyiv Gestalt University in a Time of War

APA (American Psychological Association). (n.d.). Resilience. In *APA Dictionary of Psychology*. Retrieved October 18, 2023, from https://dictionary.apa .org/resilience

Klein, M. (2017). *The collected works of Melanie Klein*. Karnac.

Perls, F. (1978). Finding Self through Gestalt Therapy. *Gestalt Journal I,* 1: 54-73.

Perls, F. (1992). *Gestalt therapy verbatim*. Gestalt Journal Press.

Perls, F., Hefferline, R., & Goodman, P. (1994). *Gestalt therapy: Excitement and growth in the human personality*. Gestalt Journal Press.

Schore, A. N. (2003). *Affect dysregulation and disorders of the self*. Norton.

When the Familiar Collapses: Living and Practicing as a Psychotherapist during a Time of War

Eger, E. (2018). *The choice: Embrace the possible*. Rider Press.

Moving Out of the Clinic Space: Intertwining Psyche, Community, and World

Bulkeley, K. (2008). *American dreamers: What dreams tell us about the political psychology of conservatives, liberals and everyone else.* Beacon Press.

Gillespie, S. (2019). Researching climate engagement: Collaborative conversations and consciousness change. In P. Hoggett (Ed.), *Climate psychology: On indifference to disaster* (pp. 107–127). Palgrave Macmillan.

Gillespie, S. (2022). Maturing conversations: Developing climate engagement through group dialogues. *Journal of Analytical Psychology, 67*(5) (In press).

Hillman, J., & Ventura, M. (1993). *We've had a hundred years of psychotherapy— And the world's getting worse.* HarperCollins.

Lawrence, W. G. (Ed.). (2007). *Infinite possibilities of social dreaming.* Karnac.

Marshall, J. (Ed.). (2009). *Depth psychology, disorder and climate change.* Jung Downunder Books.

Monbiot, G. (2006). *Heat: How to stop the planet burning.* Allen Lane.

Watkins, M., & Shulman, H. (2008). *Towards psychologies of liberation.* Palgrave Macmillan.

Weintrobe, S. (2021). *Psychological roots of the climate crisis: Neoliberal exceptionalism and the culture of uncare.* Bloomsbury.

Moving Out of the Consulting Room and Living with the Climate Crisis

Denborough, D. (2008). *Collective narrative practice.* Dulwich Centre Publications.

Ganz, M. (2011). Public narrative, collective action, and power. In S. Odugbemi and T. Lee (Eds.), *Accountability through public opinion: From inertia to public action* (pp. 273–289). World Bank. http://nrs.harvard.edu/urn-3:HUL.InstRepos:29314925

Ncube, N. (2017). *Tree of life practitioners guide.* Phola.

Randall, R. (2005). A new climate for psychotherapy? *Psychotherapy and Politics International, 3*: 3.

Whitaker, D. S. (1985). *Using groups to help people.* Routledge.

Worden, W. (1983). *Grief counseling and grief therapy.* Tavistock.

Moving with Storms

Andreotti, V., Arefin, R., Cardoso C., Evans, R., Slaymaker, O., Stein, S., Valley, W. S., & Wittman, H. (Eds.). (2023). *Moving with storms: Climate and nature emergency catalyst program (not your typical) report.* Peter Wall Institute for Advanced Studies.

Levin, K., Cashore, B., Bernstein, S., & Auld, G. (2012). Overcoming the tragedy of super wicked problems: Constraining our future selves to ameliorate global climate change. *Policy Sciences, 45*: 123–152.

Machado de Oliveira, V. (2021). *Hospicing modernity: Facing humanity's wrongs and the implications for social activism.* North Atlantic Books.
Rittel, H., & Webber, M. (1974). Wicked problems. *Man-made Futures, 26*(1): 272–280.

Collective Consolation: The Paradox of Climate Cafés

Hickman, C., Marks, E., Pihkala, P., Clayton, S., Lewandowski, R. E., Mayall, E. E., Wray, B., Mellor, C., & van Susteren, L. (2021). Climate anxiety in children and young people and their beliefs about government responses to climate change: A global survey. *The Lancet Planetary Health, 5*(12): e863–e873.
Hine, Dougald (2019). After we stop pretending. Retrieved August 16, 2022, from https://dougald.nu/after-we-stop-pretending
Norgaard, K. (2011). *Living in denial: Climate change, emotions and everyday life.* MIT Press.
Norgaard, K. M. (2018). The sociological imagination in a time of climate change. *Global and Planetary Change, 163*: 171–176.
Underwood, J. (2014). The death café. Retrieved May 9, 2019, from https://deathcafé.com/what
Weintrobe, S. (2020). Moral injury, the culture of uncare and the climate bubble. *Journal of Social Work Practice, 34*(4): 351–362.
Weintrobe, S. (2021). *Psychological roots of the climate crisis: Neoliberal exceptionalism and the culture of uncare.* Bloomsbury.

Social Dreaming

Freud, S. (1991). *The interpretation of dreams.* Penguin.
Keats, J. (1958). *The letters of John Keats.* Oxford University Press.
Lawrence, W. G. (2005). *Introduction to social dreaming: Transforming thinking.* Karnac.
Manley, J. (2018). *Social dreaming, associative thinking and intensities of affect.* Palgrave Macmillan.
Manley, J. (2020). The jewel in the corona: Crisis, the creativity of social dreaming, and climate change. *Journal of Social Work Practice, 34*(4): 429–443.
Morton, T. (2013). *Hyperobjects.* University of Minnesota Press.

Index

About North Atlantic Books

North Atlantic Books (NAB) is an independent, nonprofit publisher committed to a bold exploration of the relationships between mind, body, spirit, and nature. Founded in 1974, NAB aims to nurture a holistic view of the arts, sciences, humanities, and healing. To make a donation or to learn more about our books, authors, events, and newsletter, please visit www.northatlanticbooks.com.